P9-DCL-577

ε Epsilon

Focus: Fractions

Instruction Manual

By Steven P. Demme

1-888-854-MATH (6284)
www.MathUSee.com

1-888-854-MATH (6284)
www.MathUSee.com

Printed in the United States of America

Epsilon

SCOPE & SEQUENCE

Math-U-See is a complete and comprehensive K-12 math curriculum. While each book focuses on a specific theme, Math-U-See continuously reviews and integrates topics and concepts presented in previous levels.

Primer

α Alpha | Focus: Single-Digit Addition and Subtraction

β Beta | Focus: Multiple-Digit Addition and Subtraction

γ Gamma | Focus: Multiplication

δ Delta | Focus: Division

ε **Epsilon** | Focus: Fractions

ζ Zeta | Focus: Decimals and Percents

Pre-Algebra

Algebra 1

Stewardship*

Geometry

Algebra 2

Pre Calculus with Trigonometry

* *Stewardship is a biblical approach to personal finance. The requisite knowledge for this curriculum is a mastery of the four basic operations, as well as fractions, decimals, and percents. In the Math-U-See sequence these topics are thoroughly covered in Alpha through Zeta. We also recommend Pre-Algebra and Algebra 1 since over half of the lessons require some knowledge of algebra. Stewardship may be studied as a one-year math course or in conjunction with any of the secondary math levels.*

HOW TO USE

Five Minutes for Success

Welcome to Epsilon. I believe you will have a positive experience with the unique Math-U-See approach to teaching math. These first few pages explain the essence of this this methodology which has worked for thousands of students and teachers. I hope you will take five minutes and read through these steps carefully.

I am assuming your student has a thorough grasp of the four basic operations; addition, subtraction, multiplication, and division.

If you are using the program properly and still need additional help, you may contact your authorized representative, or visit Math-U-See online at http://www.mathusee.com/support.html

— S. Demme

The Goal of Math-U-See

The underlying assumption or premise of Math-U-See is that the reason we study math is to apply math in everyday situations. Our goal is to help produce confident problem solvers who enjoy the study of math. These are students who learn their math facts, rules, and formulas and are able to use this knowledge to solve word problems and real life applications. Therefore, the study of math is much more than simply committing to memory a list of facts. It includes memorization, but it also encompasses learning the underlying concepts of math that are critical to successful problem solving.

More than Memorization

Many people confuse memorization with understanding. Once while I was teaching seven junior high students, I asked how many pieces they would each receive if there were fourteen pieces. The students' response was, "What do we do: add, subtract, multiply, or divide?" Knowing how to divide is important, understanding when to divide is equally important.

THE SUGGESTED 4-STEP MATH-U-SEE APPROACH

In order to train students to be confident problem solvers, here are the four steps that I suggest you use to get the most from the Math-U-See curriculum.

Step 1. Prepare for the Lesson
Step 2. Present the New Topic
Step 3. Practice for Mastery
Step 4. Progression after Mastery

Step 1. Prepare for the Lesson.

Watch the DVD to learn the new concept and see how to demonstrate this concept with the manipulatives when applicable. Study the written explanations and examples in the instruction manual. Many students watch the DVD along with their instructor.

Step 2. Present the New Topic

Present the new concept to your student. Have the student watch the DVD with you, if you think it would be helpful. Older students may watch the DVD on their own.

a. Build: Use the fractions to demonstrate the problems from the worksheet.

b. Write: Record the step-by-step solutions on paper as you work them through with manipulatives.

c. Say: Explain the *why* and *what* of math as you build and write.

Do as many problems as you feel are necessary until the student is comfortable with the new material. One of the joys of teaching is hearing a student say *"Now I get it!"* or *"Now I see it!"*

Step 3. Practice for Mastery.

Using the examples and the lesson practice problems from the student text, have the students practice the new concept until they understand it. It is one thing for students to watch someone else do a problem, it is quite another to do the same problem themselves. Do enough examples together until they can do them without assistance.

Do as many of the lesson practice pages as necessary (not all pages may be needed) until the students remember the new material and gain understanding. Give special attention to the word problems, which are designed to apply the concept being taught in the lesson.

Another resource is the Math-U-See web site which has online drill and downloadable worksheets for several of the lessons. Click on www.mathusee.com and select "Online Helps."

Step 4. Progression after Mastery.

Once mastery of the new concept is demonstrated, proceed to the systematic review pages for that lesson. Mastery can be demonstrated by having each student teach the new material back to you. The goal is not to fill in worksheets, but to be able to teach back what has been learned.

The systematic review worksheets review the new material as well as provide practice of the math concepts previously studied. Remediate missed problems as they arise to ensure continued mastery.

Proceed to the lesson tests. These were designed to be an assessment tool to help determine mastery, but they may also be used as an extra worksheet. Your students will be ready for the next lesson only after demonstrating mastery of the new concept and continued mastery of concepts found in the systematic review worksheets.

Confucius was reputed to have said, "Tell me, I forget; Show me, I understand; Let me do it, I will remember." To which we add, **"Let me teach it and I will have achieved mastery!"**

Length of a Lesson

So how long should a lesson take? This will vary from student to student and from topic to topic. You may spend a day on a new topic, or you may spend several days. There are so many factors that influence this process that it is impossible to predict the length of time from one lesson to another. I have spent three days on a lesson and I have also invested three weeks in a lesson. This occurred in the same book with the same student. If you move from lesson to lesson too quickly without the student demonstrating mastery, he will become overwhelmed and discouraged as he is exposed to more new material without having learned the previous topics. But if you move too slowly, your student may become bored and lose interest in math. But I believe that as you regularly spend time working along with your student, you will sense when is the right time to take the lesson test and progress through the book.

By following the four steps outlined above, you will have a much greater opportunity to succeed. Math must be taught sequentially, as it builds line upon line and precept upon precept on previously learned material. I hope you will try this methodology and move at your student's pace. As you do, I think you will be helping to create a confident problem solver who enjoys the study of math.

ONGOING SUPPORT AND ADDITIONAL RESOURCES

Welcome to the Math-U-See Family!

Now that you have invested in your children's education, I would like to tell you about the resources that are available to you. Allow me to introduce you to your regional representative, our ever improving website, the Math-U-See blog, our new free e-mail newsletter, the online Forum, and the Users Group.

Most of our regional **Representatives** have been with us for over 10 years. What makes them unique is their desire to serve and their expertise. They have all used Math-U-See and are able to answer most of your questions, place your student(s) in the appropriate level, and provide knowledgeable support throughout the school year. They are wonderful!

Come to your local curriculum fair where you can meet your rep face-to-face, see the latest products, attend a workshop, meet other MUS users at the booth, and be refreshed. We are at most curriculum fairs and events. To find the fair nearest you, click on "Events Calendar" under "News."

The **Website**, at www.mathusee.com, is continually being updated and improved. It has many excellent tools to enhance your teaching and provide more practice for your student(s).

2
+ 1

ONLINE DRILL

Let your students review their math facts online. Just enter the facts you want to learn and start drilling. This is a great way to commit those facts to memory.

WORKSHEET GENERATOR

Create custom worksheets to print out and use with your students. It's easy to use and gives you the flexibility to focus on a specific lesson. Best of all — it's free!

Math-U-See Blog

Interesting insights and up-to-date information appear regularly on the Math-U-See Blog. The blog features updates, rep highlights, fun pictures, and stories from other users. Visit us and get the latest scoop on what is happening.

Email Newsletter

For the latest news and practical teaching tips, sign up online for the free Math-U-See e-mail newsletter. Each month you will receive an e-mail with a teaching tip from Steve as well as the latest news from the website. It's short, beneficial, and fun. Sign up today!

The Math-U-See Forum and the Users Group put the combined wisdom of several thousand of your peers with years of teaching experience at your disposal.

Online Forum

Have a question, a great idea, or just want to chitchat with other Math-U-See users? Go to the online forum. You can also use the forum to post a specific math question if you are having difficulty in a certain lesson. Head on over to the forum and join in the discussion.

Yahoo Users Group

The MUS-users group was started in 1998 for lovers and users of the Math-U-See program. It was founded by two home-educating mothers and users of Math-U-See. The backbone of information and support is provided by several thousand fellow MUS users.

For Specific Math Help

When you have watched the DVD instruction and read the instruction manual and still have a question, we are here to help. Call your local rep, click the support link and e-mail us here at the home office, or post your question on the forum. Our trained staff has used Math-U-See themselves and are available to answer a question or walk you through a specific lesson.

Feedback

Send us an e-mail by clicking the feedback link. We are here to serve you and help you teach math. Ask a question, leave a comment, or tell us how you and your student are doing with Math-U-See.

Our hope and prayer is that you and your students will be equipped to have a successful experience with math!

Blessings,

Steve

Steve Demme

LESSON 1

Fraction of a Number

Word Problem Tips

There are three steps to help you understand fractions. Work on these steps in this lesson, and we will use them as a basis for much of the book. If you have the Math-U-See blocks, you may use the unit blocks for this lesson. If you don't have the blocks, you may use any individual objects like raisins or pieces of candy.

The first step is often left out of most teaching about fractions. It is very important because it is what you begin with. A fraction is a fraction of something. This lesson comes first to magnify this fact. I often ask students, "What is larger, one-half or one-fourth?" They usually reply, "One-half," when what they should say is, "One-half of what and one-fourth of what?" If I then say, "One-half of the room we are in, or one-fourth of the state?" their answer will be different. So, the first step is knowing what number you are starting with.

The second step is to look at what you divide the number into, which is called the *denominator*. There is a natural language connection since both divide and denominator begin with D. The symbolism is also an indicator of this step since the line separating the numerator and denominator means "divided by." So you divide step 1 by the bottom number of the fraction, the denominator, into that many equal parts.

In the third step you count how many equal parts. I call this the "numBerator" at first to make the connection with counting. But after a while we take out the B and have *numerator*. To summarize, here are the three points:

1. The number you start with; in example 1 below, this is six.
2. The denominator, which indicates the "value" (as in place value), and tells how many equal parts we divide the starting number into.
3. The numBerator or numerator, which is "how many" of these equal parts we count.

Example 1

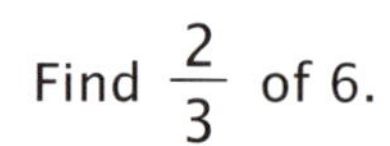

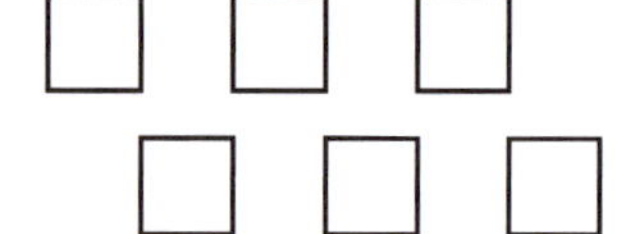

Step 1 - Select 6 green unit blocks.

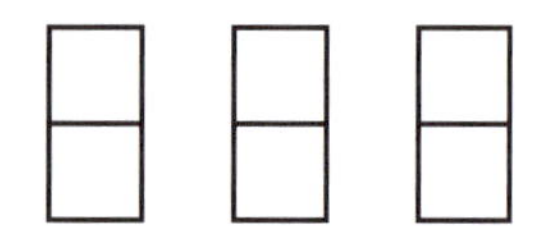

Step 2 - Divide 6 into 3 equal parts.

$\frac{2}{3}$ of 6 = 4

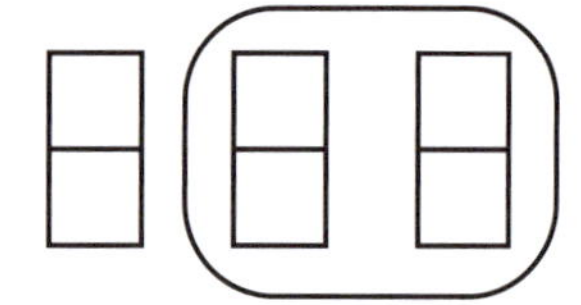

Step 3 - Count 2 of those parts.

A *fraction* of a number is a combination of a division problem and a multiplication problem. First you are dividing (in the example, by three for the equal parts) to get the denominator, then multiplying (by two) to get the numerator. $6 \div 3 \times 2 = 4$.

Example 2

Find $\frac{3}{4}$ of12.

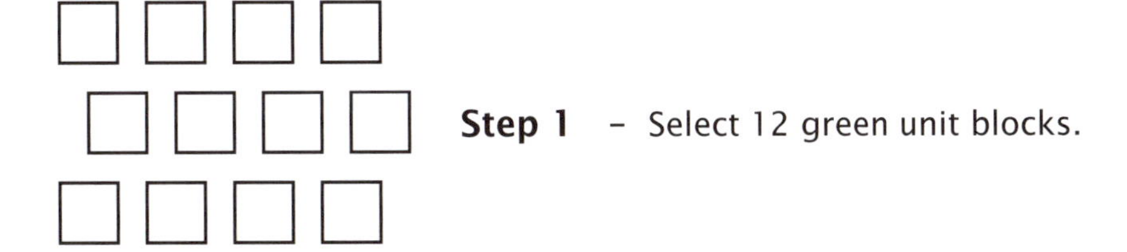

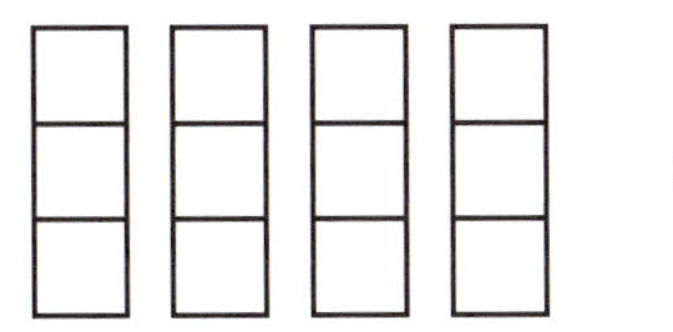

Step 2 - Divide 12 into 4 equal parts.

$\frac{3}{4}$ of 12 = 9

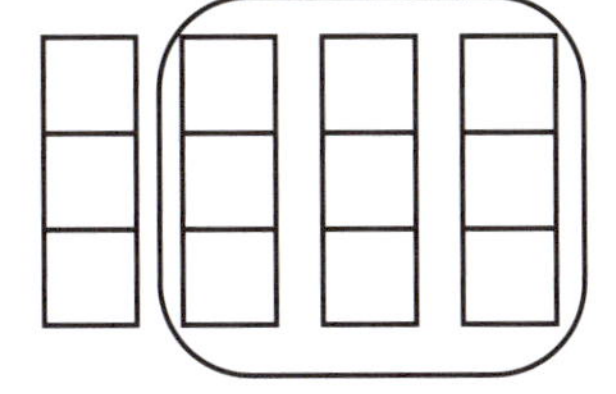

Step 3 - Count 3 of those parts.

WORD PROBLEM TIPS

Parents often find it challenging to teach children how to solve word problems. Here are some suggestions for helping your student learn this important skill.
The first step is to realize that word problems require both reading and math comprehension. Don't expect a child to be able to solve a word problem if he does not thoroughly understand the math concepts involved. On the other hand, a student may have a math skill level that is stronger than his or her reading comprehension skills. Below are a number of strategies to improve comprehension skills in the context of story problems. You may decide which ones work best for you and your child.

STRATEGIES FOR WORD PROBLEMS

1. Ignore numbers at first and read the story. It may help some students to read the question aloud. Every word problem tells a story. Before deciding what math operation is required, let the student retell the story in his own words. Who is involved? Are they receiving gifts, losing something, or dividing a treat?
2. Relate the story to real life, perhaps by using names of family members. For some students, this makes the problem more interesting and relevant.
3. Build, draw, or act out the story. Use the blocks or actual objects when practical. Especially in the lower levels, you may require the student to use the blocks for word problems even when the facts have been learned. Don't be afraid to use a little drama as well. The purpose is to make it as real and meaningful as possible.
4. Look for the common language used in a particular kind of problem. Pay close attention to the word problems on the lesson practice pages, as they model different kinds of language that may be used for the new concept just studied. For example, "altogether" indicates addition. These "key words" can be useful clues but should not be a substitute for understanding.
5. Look for practical applications that use the concept and ask questions in that context.
6. Have the student invent word problems to illustrate his number problems from the lesson.

CAUTIONS

1. Unneeded information may be included in the problem. For example, we may be told that Suzie is eight years old, but the eight is irrelevant when adding up the number of gifts she received.
2. Some problems may require more than one step to solve. Model these questions carefully.
3. There may be more than one way to solve some problems. Experience will help the student choose the easier or preferred method.
4. Estimation is a valuable tool for checking an answer. If an answer is unreasonable, it is possible that the wrong method was used to solve the problem.

Note: Lessons 9 and 10 in the *Epsilon* student text have some tips for fraction word problems.

LESSON 2

Fraction of One

Instead of a fraction of a number, we'll talk about a fraction of one in this lesson. I've been using the green unit block to represent one. To transition from the blocks to the overlays, I select the green 5 x 5 inch square to represent one. Then I hold up a white square and place it over the green one, showing that they are the same size and thus the same value—one. I prefer white as we proceed, since it has some value but is not a color to be confused with the other color inserts.

Now, holding the white square as one (the first step, or what we start with), take the clear overlay with four vertical lines, which when placed on top of the one, divides it into five equal parts (denominator). We've done two steps so far. We've taken one and divided it into five equal parts. Now what is the number of those equal parts? Or, how many of those equal parts are we going to count (numerator)? In this case the number is two, and the light blue colored piece is the numerator.

Example 1

Find $\frac{2}{5}$ of 1.

Step 1 - Begin with 1.

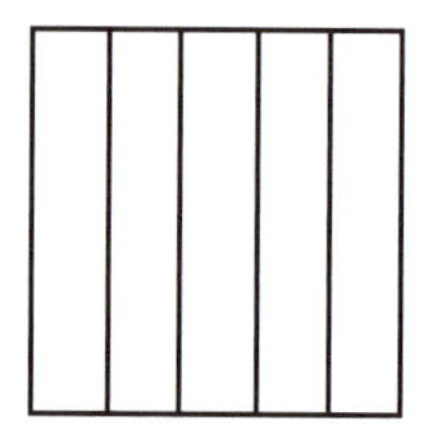

Step 2 - Divide 1 into 5 equal parts.

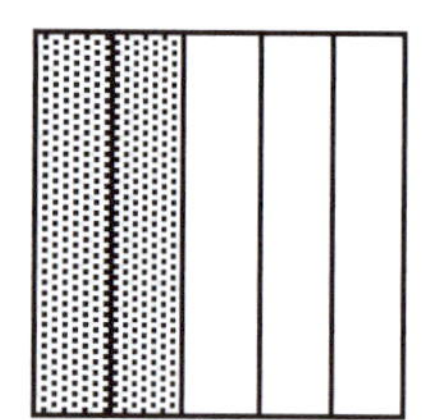

Step 3 - Count 2 of those parts.

Example 2

Find $\frac{3}{4}$ of 1.

Step 1 - Begin with 1.

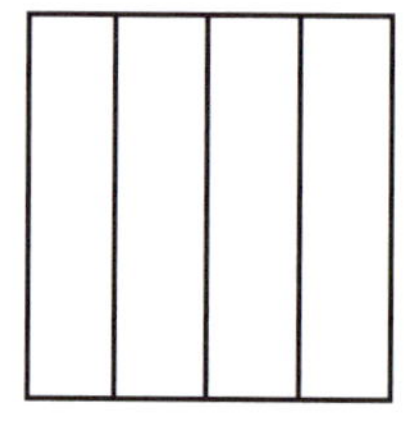

Step 2 - Divide 1 into 4 equal parts.

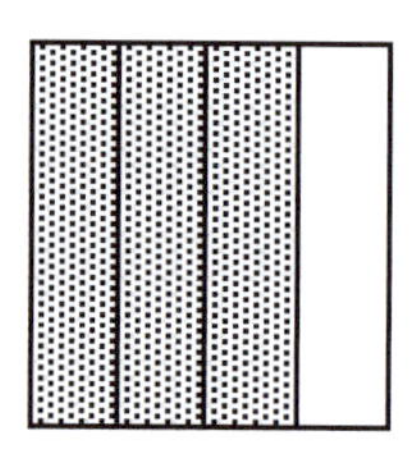

Step 3 - Count 3 of those parts.

These are the same steps used to find a fraction of a number. Remember that the horizontal line separating the numerator and the denominator means "divided by." Before you begin the worksheets read the following instructions carefully.

A. Build several different fractions with the overlays and ask the student to write what you've built. Then have him read the correct answer out loud. When the student is comfortable with this, reverse the roles and ask the student to build several for you to write out, then say your answers.

B. When you feel the student grasps this concept, write a fraction and have him build it, and then say what he has built. Do this again with the student as the instructor.

C. Now say a fraction out loud and have the student build and write it. Then have the student say one and you build and write it.

Use this process with all new material whenever possible even if the directions in each section do not specifically tell you to. Build—Write—Say, and then use role reversal to model good problem-solving skills for the student.

LESSON 3

Add and Subtract Fractions

with the Same Denominator; Mental Math

To add or subtract fractions, you must have the same number of equal parts. As in place value, you can add only tens to tens, units to units, or hundreds to hundreds. Two numbers may be combined only if they are the same value. In fractions, the denominator indicates what kind or value. In order to add two fractions, the denominators must be the same. This is our unit of comparison, or unit of measure, or value, and is referred to as the *common denominator*. I usually say "same denominator" for some time before I use the words, "common denominator." This makes more sense to the student and makes the concept clearer.

Let's illustrate this by adding 1/4 to 2/4. Make 1/4 with the overlays. Beginning with the white one, place the fourths overlay on top and slide the thin yellow strip between. Now take the other clear overlay in the fourths pocket and place it on top of the yellow piece that represents 2/4, without a white background. Now place the 2/4 on top of the 1/4. You can see that the answer is 3/4 as everything lines up. Do several of these until you understand that when you have the same denominator you add the numerators. The same holds true for subtraction.

Example 1

Solve $\frac{2}{4} + \frac{1}{4} =$

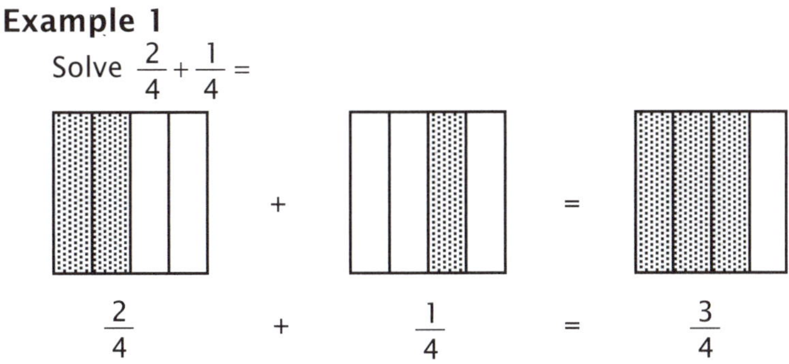

$\frac{2}{4} + \frac{1}{4} = \frac{3}{4}$

Two-fourths plus one-fourth equals three-fourths.

Example 2

Solve $\frac{2}{6}+\frac{3}{6}=$

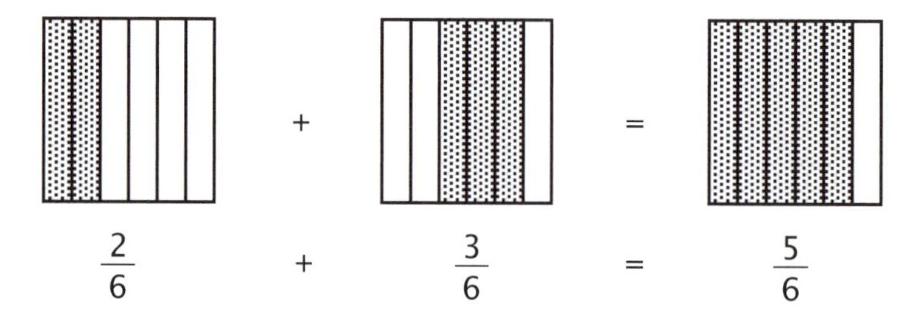

Two-sixths plus three-sixths equals five-sixths.

Example 3

Solve $\frac{5}{6}-\frac{1}{6}=$

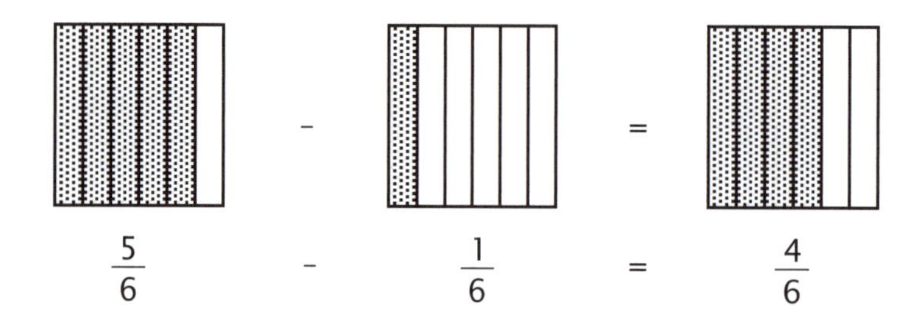

Five-sixths minus one-sixth equals four-sixths.

MENTAL MATH

Mental math problems can be used to keep the facts alive in the memory and to develop mental math skills. Read the problem slowly enough so that the student comprehends, and then walks him through increasingly difficult exercises. The purpose is to stretch but not discourage. You decide where that line is!

Example 4

Five times three, plus six, divided by seven, equals?

The student thinks: 5 x 3 = 15, 15 + 6 = 21, and 21 ÷ 7 = 3.

At first you will need to go slowly enough for the student to verbalize the intermediate steps. As skills increase, the student should be able to give just the answer. Notice that the example includes an addition problem that is not a basic addition fact.

These questions include addition, subtraction, multiplication, and division. Do a few at a time, and go slowly at first.

1. Three plus five, times six, minus three, equals ? (45)
2. Eight times three, divided by six, plus nine, equals ? (13)
3. Nineteen minus nine, plus 10, divided by five, equals ? (4)
4. Twenty-seven divided by three, times five, plus two, equals ? (47)
5. Five plus four, times eight, plus six, equals ? (78)
6. Fifty-four divided by six, divided by three, times seven, equals ? (21)
7. Thirty-two minus two, times three, divided by 10, equals ? (9)
8. Seven minus five, times eight, divided by four, equals ? (4)
9. Seven times seven, plus one, divided by five, equals ? (10)
10. Nine plus two, times four, minus five, equals ? (39)

There are more mental math problems in lessons 9, 15, and 21. There are also mental math problems in the student text, starting with lesson 21D.

LESSON 4

Equivalent Fractions

The next three sections in fractions—adding, subtracting, and dividing fractions, as well as reducing fractions—all hinge on equivalent fractions. I call this the watershed of fractions. You must understand this concept to thoroughly understand fractions. On paper, it is difficult to comprehend how 1/2 could be the same as 2/4 and 3/6 and 4/8. Look through the fractions kit now. Notice the colors are synchronized with the blocks. Orange is the color for the two unit bar as well as the halves. Pink is for three in the blocks and thirds in the fraction kit, and so on. Begin by making 1/2. Use the white piece for your background, then place the clear overlay with one black line through it vertically over the piece, then finally, place the small orange half piece between these two pieces. We begin with one (white piece), divide it, or cut it up, into two pieces (clear overlay), and count one of them (orange piece). After you have made 1/2, then place the other clear overlay found in the halves pocket horizontally on top of this to show 2/4. Take this piece off, and place the clear overlay from the thirds pocket horizontally on top to show 3/6. Now, try the fourths and fifths overlays.

Example 1

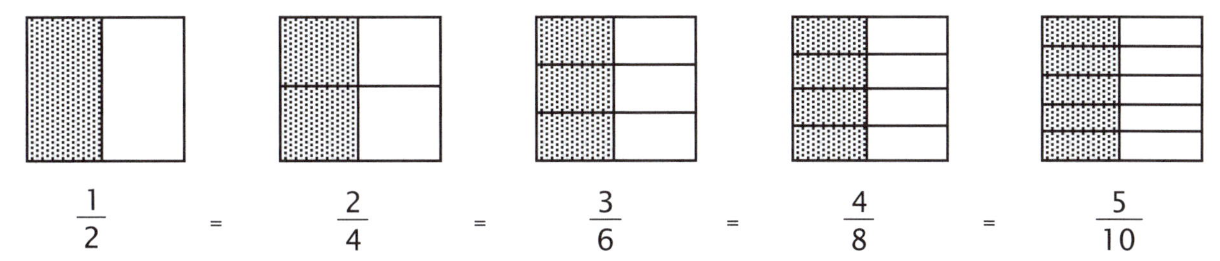

One-half equals two-fourths equals three-sixths equals four-eighths equals five-tenths

At this point I tell a story, and then ask a question. Let's say you have one-half of a pizza left over from dinner, which you were planning to eat for lunch. Then the doorbell rings and two friends drop by. You invite them in, cut the pizza into three equal parts, and you each get one piece. After you cut the pizza, and before you give it to your friends, stop and think. Do you still have the same amount of pizza? Yes. But are there more pieces? Yes. What fraction did I have originally? One-half. Now what do I have (with the overlay added)? Three-sixths. Do I still have the same amount? Yes. (Take off the second overlay if needed to show that it is still one-half). But do I have more pieces (or equal parts)? Yes. Keep trying different overlays and combinations until you understand this critical concept, which I summarize by saying, "Same amount, more pieces."

Example 2

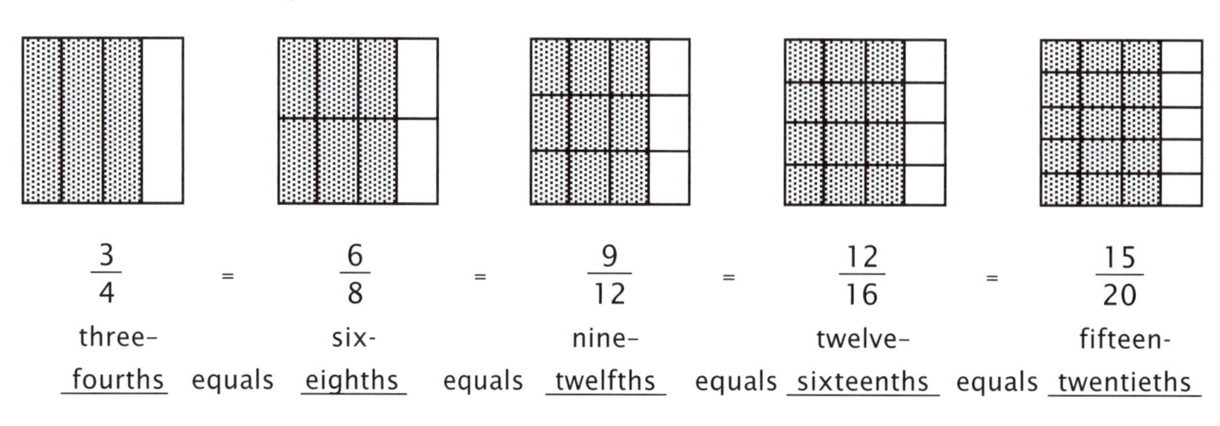

Example 3

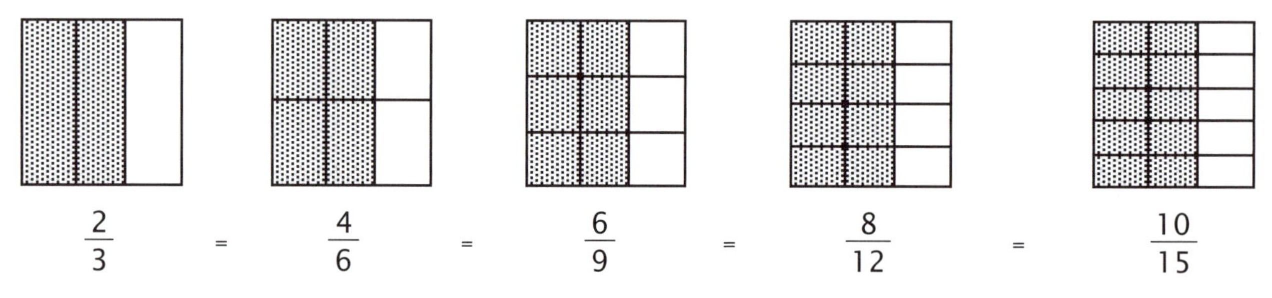

two-thirds equals four-sixths equals six-ninths equals eight-twelfths equals ten-fifteenths

LESSON 5

Addition and Subtraction
with Unequal Denominators

When adding fractions with unequal denominators, please do it the long way as mentioned in the instructions and as shown on the DVD. Read this section carefully, take your time, and use the overlays.

To add or subtract fractions, you must have the same number of equal parts. As in place value, you can add only tens to tens, units to units, or hundreds to hundreds. So, in fractions, the denominators must be the same. This is our unit of comparison, or unit of measure, or our value. We call this a common denominator. We have been taking this concept into account while adding fractions like 1/5 and 3/5. We understand that when we have the same denominator, we add the numerators. The same holds true for subtraction. Again, I usually say, "same denominator," for some time before I use the words, "common denominator." This usually makes more sense to the student and makes the concept clearer.

Now try adding 2/5 and 1/3 with the overlays the same as you did with 1/5 plus 3/5. When you place one on top of the other, you can see that there is no clear answer. It is a mess. We cannot combine them unless they have the same denominator or are the same kind. So in order to add or subtract two fractions, we must first find the same denominator.

But after the concept of equivalent fractions has been mastered, this kind of problem is not so tough, because we know how to change the number of pieces while still keeping the same amount. What we want to do then is to keep changing the 2/5 and the 1/3 until they have the same denominator. Remember our formula that we discovered: if the denominators are the same, we can combine the numerators.

So we start by making the two fractions, in this case 1/3 and 2/5. Then we begin placing the overlays, beginning with the 1/2 overlay, on top of the fractions to be added. This doesn't give us the same denominator, so we try the 1/3 overlay. We keep trying overlays until we find the same denominator. Then we add the numerators. Study the two examples.

Example 1

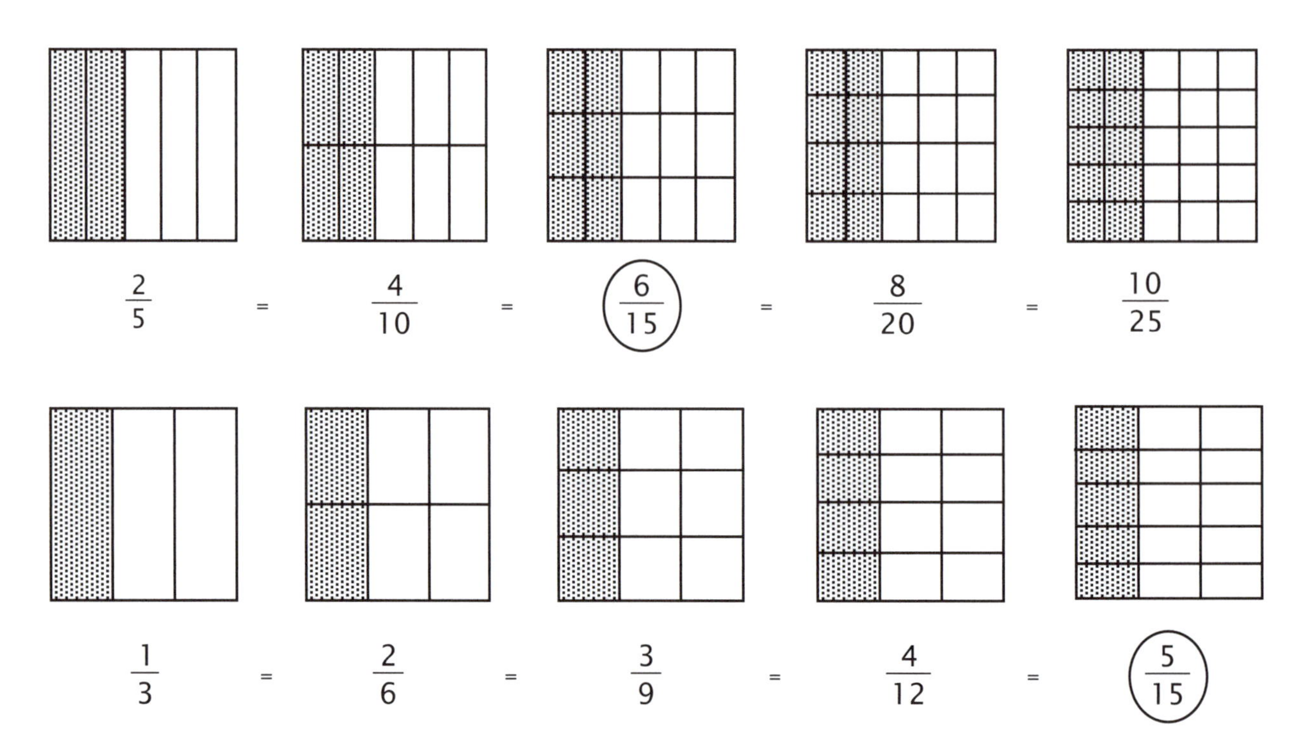

So 6/15 + 5/15 is 11/15. Notice that the size of the little rectangles is the same. Turn one of the fractions 90 degrees so that you can see that each rectangle is exactly the same size.

Example 2

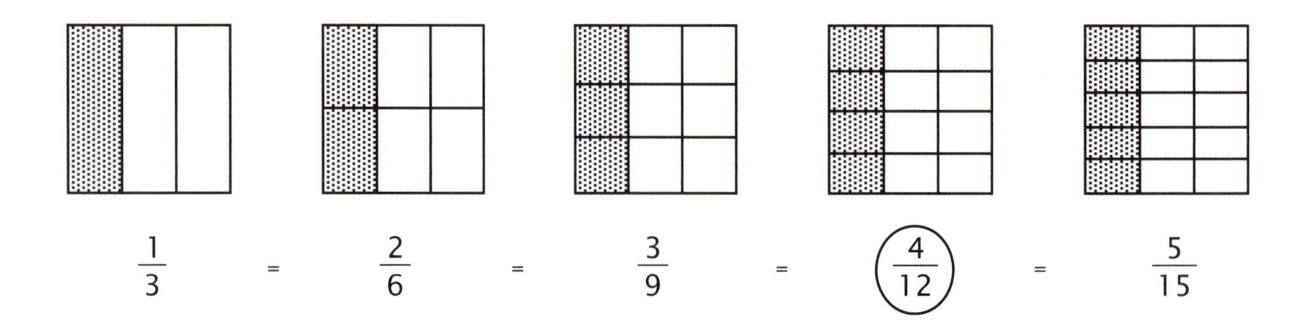

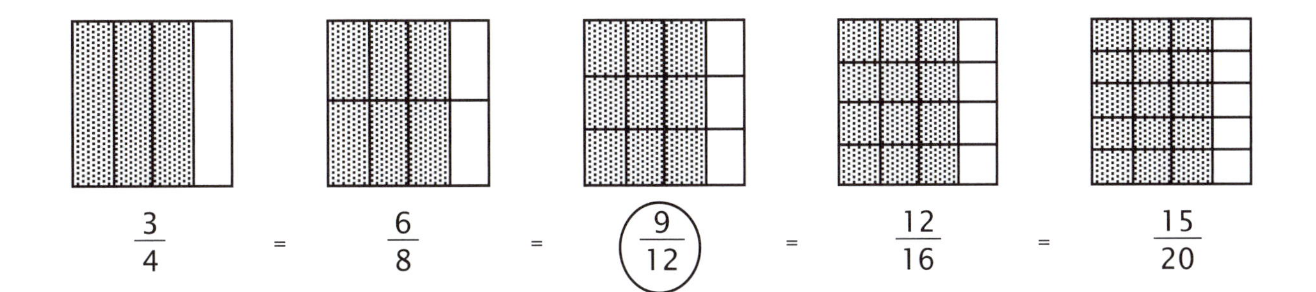

$\frac{4}{12} + \frac{9}{12} = \frac{13}{12}$ Notice that 13/12 is more than 12/12,or one. We will learn more about this kind of fraction later.

After you do this for several of the worksheets, the students should begin to observe the pattern: that to get the same denominator, all they need to do is crisscross the denominator overlays.

Example 2 (with the shortcut)

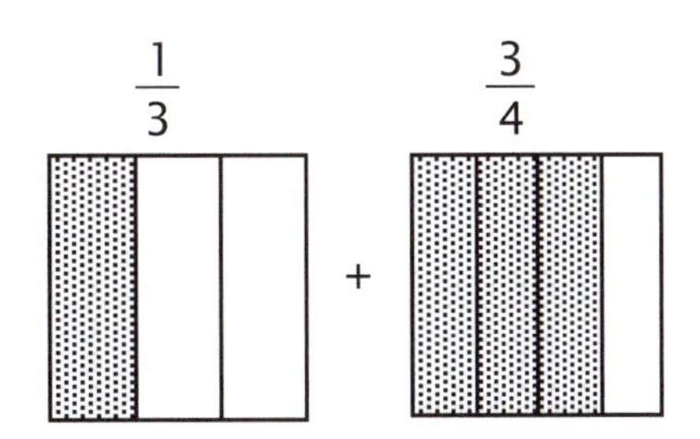

While holding 1/3, take the other fourth overlay and place it sideways on 1/3.

While holding 3/4, take the other third overlay and place it sideways on 3/4.

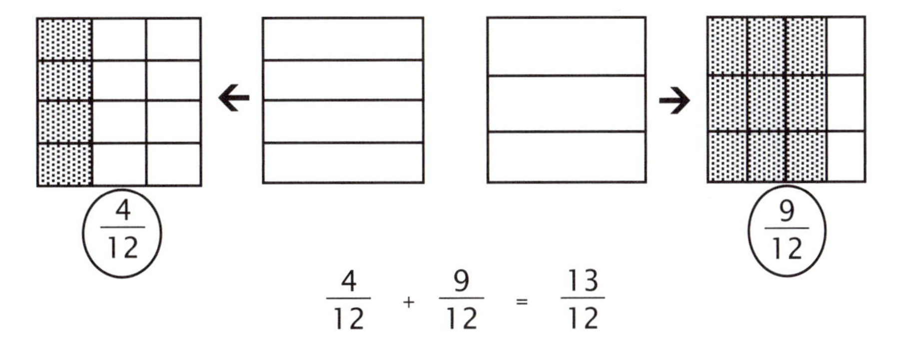

This method will always work. It will not always give you the least common denominator, but it will always give you a common denominator. This shortcut is shown in greater detail in the next lesson.

Example 1 (with the shortcut)

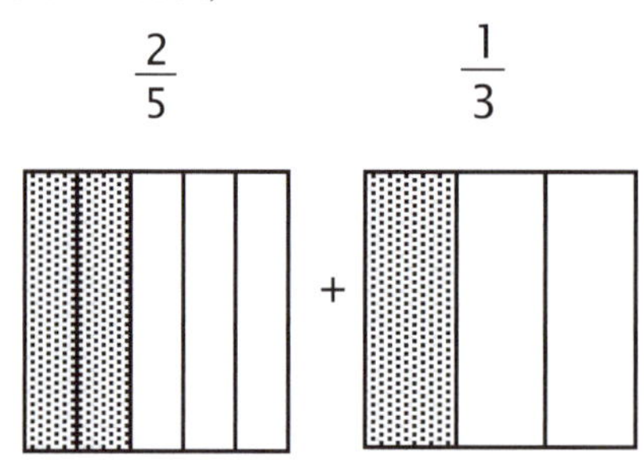

While holding 2/5, take the other third overlay and place it sideways on 2/5.

While holding 1/3, take the other fifth overlay and place it sideways on 1/3.

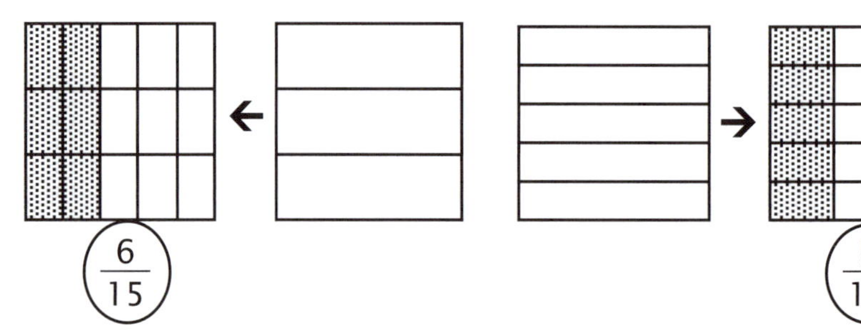

$$\frac{6}{15} + \frac{5}{15} = \frac{11}{15}$$

LESSON 6

The Rule of Four

Multi-step Word Problems

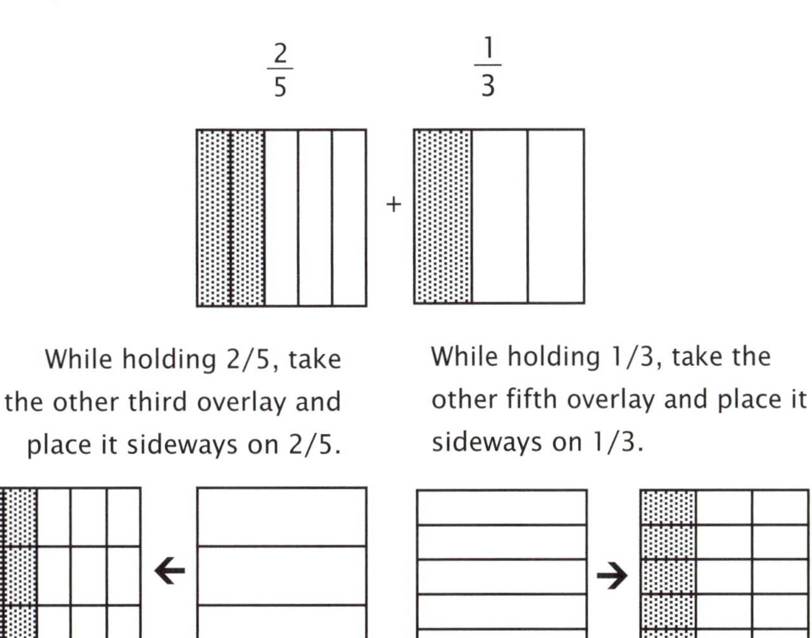

The rule of four is the formula for the shortcut we were doing at the end of the last lesson. The diagram shows what happens when you crisscross overlays. Steps 1 and 2 are a result of placing the thirds overlay on top of 2/5. The numerator and denominator are both increased by a factor of three. Steps 3 and 4 result from placing the fifths overlay on top of 1/3. The numerator and denominator are both increased by a factor of five.

Rule of Four - For adding, subtracting, comparing, and dividing fractions. The four denotes the four steps.

Example 1

Step 1	3 x 5 = 15
Step 2	3 x 2 = 6
Step 3	5 x 3 = 15
Step 4	5 x 1 = 5

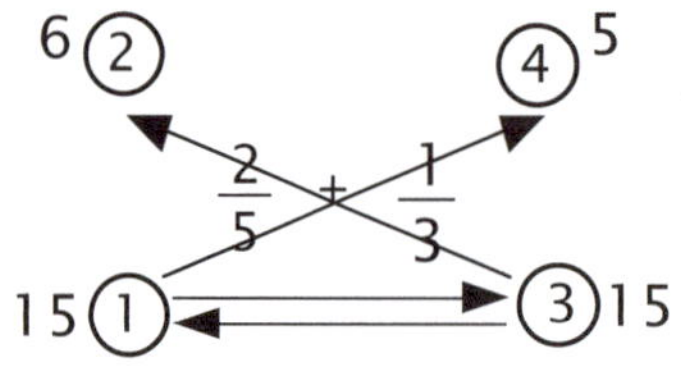

$$\frac{2}{5} + \frac{1}{3}$$

$$\frac{6}{15} + \frac{5}{15} = \frac{11}{15}$$

Remember, this rule always works. It will not always give you the least common denominator, but it will always give you a common denominator.

Example 2

Step 1	6 x 4 = 24
Step 2	6 x 3 = 18
Step 3	4 x 6 = 24
Step 4	4 x 1 = 4

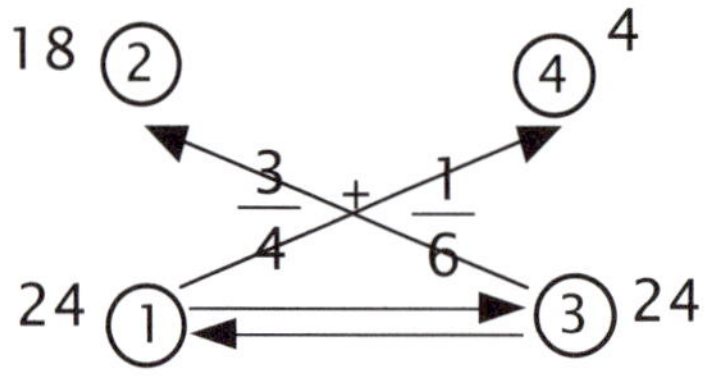

$$\frac{3}{4} + \frac{1}{6}$$

$$\frac{18}{24} + \frac{4}{24} = \frac{22}{24}$$

Example 3

Step 1	7 x 8 = 56
Step 2	7 x 5 = 35
Step 3	8 x 7 = 56
Step 4	8 x 2 = 16

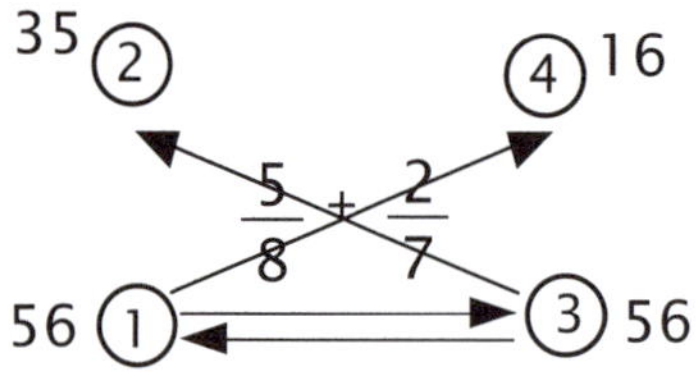

$$\frac{5}{8} + \frac{2}{7}$$

$$\frac{35}{56} + \frac{16}{56} = \frac{51}{56}$$

MULTISTEP WORD PROBLEMS

The student text includes some fairly simple two-step word problems. Some students may be ready for more challenging problems. Here are a few to try, along with some tips for solving this kind of problem. You may want to read and discuss these with your student as you work out the solutions together. The purpose is to stretch, not to frustrate. If you do not think the student is ready, you may want to come back to these later.

There are more multistep word problems in lessons 12, 18, and 24 in this book. The answers are at the end of the solutions section at the back of the book.

1. Tom planted vegetables in a rectangular garden that was 25 feet long and 15 feet wide. He used one-third of the area for corn and one-fifth of it for peas. How many square feet are left for other vegetables?

Although the problem asks only one question, there are other questions that must be answered first. The key to solving this problem is determining what the unstated questions are. Since the final question is asking for the leftover area, the unstated questions are "What is the total area of Tom's garden?" and "What is the area used for each of the vegetables mentioned?"

You might make a list of questions something like this:

A. Area of garden in square feet?
B. Area used for corn?
C. Area used for peas?
D. Total area used for peas and corn?
E. Leftover area?

2. Sarah signed one-half of the Christmas cards, and Richard signed three-eighths of them. If there are 32 cards in all, how many are left to be signed?

3. Jim wishes to buy three gifts that cost $15, $9, and $12. He has one-fourth of the money he needs. How much more money must he earn in order to buy the gifts?

LESSON 7

Comparing Fractions with the Rule of Four

So far we know that “=” means “equals” or “is the same as.” If two fractions are not equal, and one is larger or smaller than the other, there are symbols to represent this. As you read an equation from left to right, “>” means “is greater than” and “<” means “is less than.” We call these symbols *inequalities*. For example, nine is greater than three, or $9 > 3$. If it were the other way around, you would write three is less than nine, or $3 < 9$. There are other ways to think of these symbols. Some say the open, or large, end of the symbols always points to the larger one, and the small end, or point, points at the smaller one. Some children think of the symbol as a hungry alligator with his mouth open, always trying to eat the larger number. Use whatever helps you remember this symbol.

Inequalities apply to fractions in two ways: denominators that are the same, and denominators that are different. If the denominators of two fractions are the same, then you just compare the numerators. An example of this is $3/4 > 1/4$, or three-fourths is greater than one-fourth. But what about 2/3 and 3/5? Which is larger? Remember what we did when adding two fractions with different denominators? We first made them the same kind, or same denominators, then combined the numerators. To compare or combine, they must be the same kind. So now we must make them the same kind, or denominator, then compare the numerators. We'll do this in the picture below. Using the overlays, we get a pretty good idea which symbol to use, but the rule of four will determine the answer.

Example 1

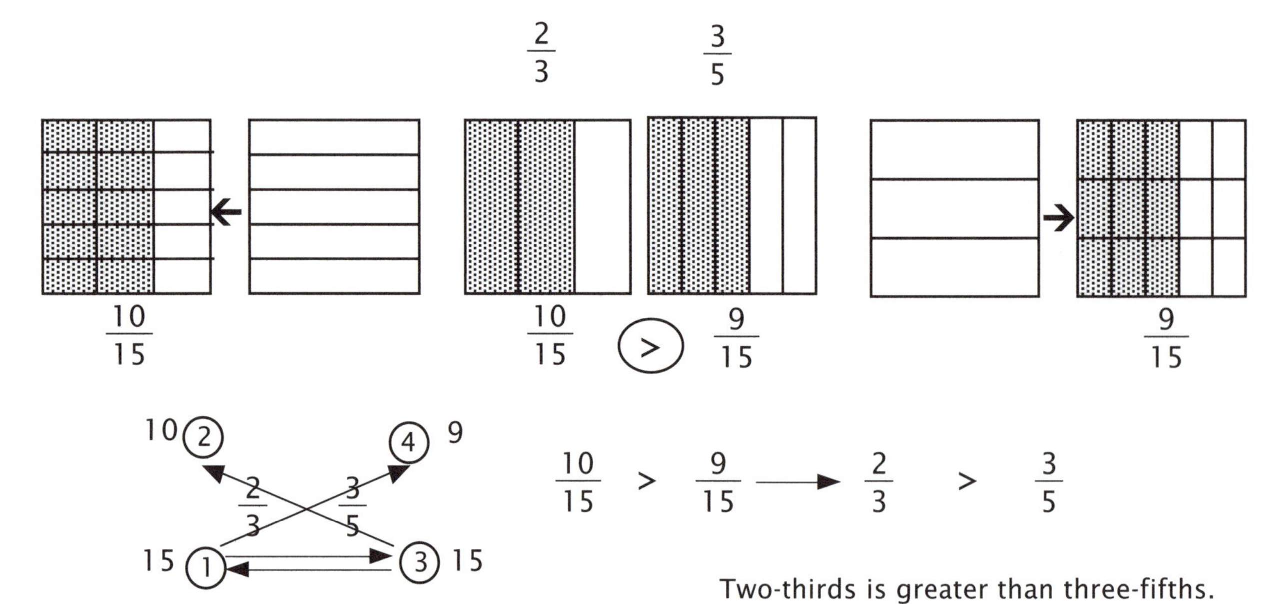

Two-thirds is greater than three-fifths.

Example 2

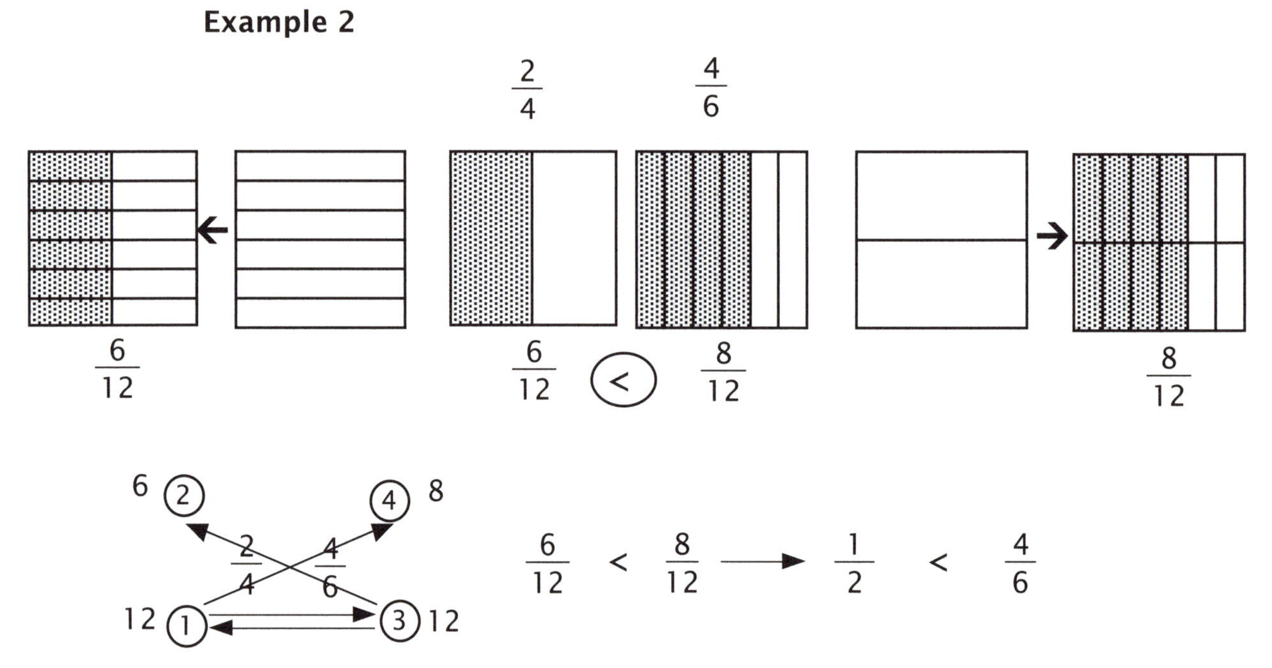

One-half is less than four-sixths.

LESSON 8

Adding Multiple Fractions

There are three methods used to find the sum of three or more fractions. Get comfortable with the first method before exploring the second and third methods. You may want to spend a day or two on each method to avoid confusion. Examples 1 and 2 are done with the first method.

Method 1–Simply add two fractions at a time using the rule of four. In example 1, add 1/2 to 3/5, then take this answer and add it to 4/7. (See note below.)

Example 1

$$\frac{1}{2}+\frac{3}{5}+\frac{4}{7} \rightarrow \frac{1}{2}+\frac{3}{5}=\frac{5}{10}+\frac{6}{10} \rightarrow \frac{11}{10}+\frac{4}{7}=\frac{77}{70}+\frac{40}{70}=\frac{117}{70}$$

Example 2

$$\frac{1}{2}+\frac{2}{3}+\frac{5}{6} \rightarrow \frac{1}{2}+\frac{2}{3}=\frac{3}{6}+\frac{4}{6} \rightarrow \frac{7}{6}+\frac{5}{6}=\frac{12}{6}=2$$

Note: Both example 1 and example 2 yield a fraction greater than one. If you remember that the line in a fraction means division, you can divide the numerator by the denominator to get a simpler answer, as in example 2 above. If the division does not come out even, write your remainder as a fraction. For now, solutions will be given in the un-simplified and simplified forms. Either form is correct.

Method 2–This method is a modification of the rule of four. Multiply the first fraction by the denominators of the second and third fractions. Multiply the second fraction by the denominators of the first and third fractions. Multiply the third fraction by the denominators of the first and second fractions. Then combine the numerators.

Example 3

$$\frac{1}{2}+\frac{3}{5}+\frac{4}{7}=\frac{1\times5\times7}{2\times5\times7}+\frac{3\times2\times7}{5\times2\times7}+\frac{4\times2\times5}{7\times2\times5}=\frac{35}{70}+\frac{42}{70}+\frac{40}{70}=\frac{117}{70}$$

Example 4

$$\frac{1}{2}+\frac{2}{3}+\frac{5}{6}=\frac{1\times3\times6}{2\times3\times6}+\frac{2\times2\times6}{3\times2\times6}+\frac{5\times2\times3}{6\times2\times3}=\frac{18}{36}+\frac{24}{36}+\frac{30}{36}=\frac{72}{36}=2$$

Method 3–This is a shorter way of doing method 2. Looking at the denominators, you can see that six could be the common denominator. So you only need to multiply each fraction by whatever number it takes to make the common denominator six for each fraction. One-half is multiplied by a factor of three. Two-thirds is multiplied by a factor of two. And five-sixths already has a denominator of six, so it is left alone. This method would not be a shortcut for example 1, since none of the denominators (2, 5, or 7) would qualify as a common denominator.

Example 5

$$\frac{1}{2}+\frac{2}{3}+\frac{5}{6}=\frac{1\times3}{2\times3}+\frac{2\times2}{3\times2}+\frac{5}{6}=\frac{3}{6}+\frac{4}{6}+\frac{5}{6}=\frac{12}{6}=2$$

LESSON 9

Multiplying Fractions

or a Fraction of a Fraction; Mental Math

We've worked on a fraction of one and a fraction of a number; now we'll tackle a fraction of a fraction. This kind of problem can be the hardest to understand or think through, but it is the easiest to do using a formula. Find 2/5 of 1/3. Notice that we don't say, "two-fifths **times** one-third." Even though it is the same as multiplying two fractions, I want to relate it to a fraction of a number and a fraction of one. To stress this relationship, we read it "two-fifths **of** one-third". This is also the language used in most word problems. Be sure to read the note about word problems on lesson practice 9A in the student text.

Example 1

Find $\frac{2}{5}$ of $\frac{1}{3}$ (two-fifths of one-third).

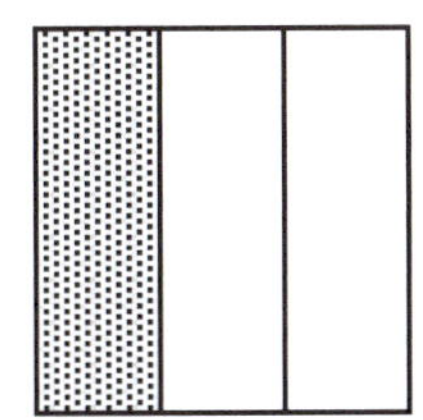

Step 1 Start with $\frac{1}{3}$.

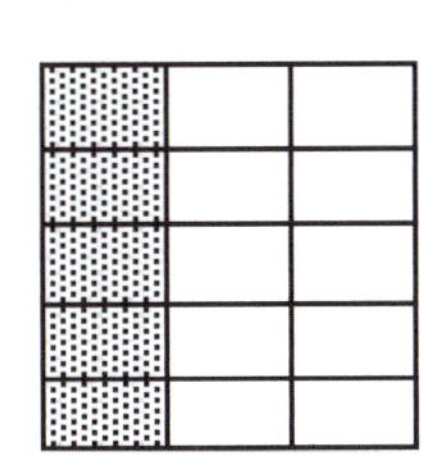

Step 2 Divide into 5 equal parts.

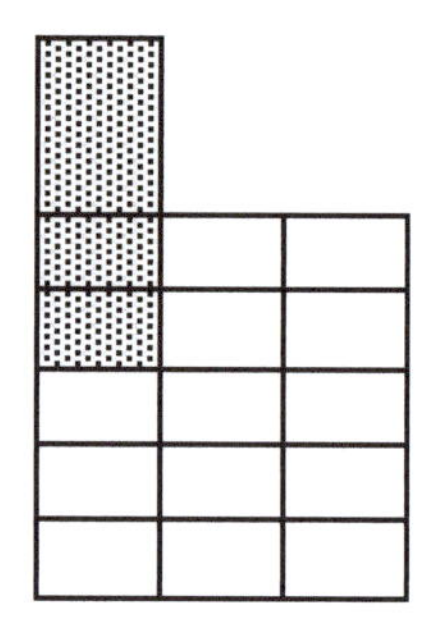

Step 3 Count 2 of those parts.

Pull up the pink insert to do this.

So $\frac{2}{5}$ of $\frac{1}{3}$ is $\frac{2}{15}$.

Example 2

Find $\frac{2}{3}$ of $\frac{5}{6}$ (two-thirds of five-sixths).

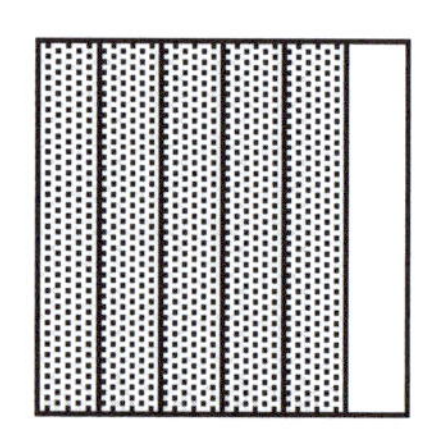

Step 1 Start with $\frac{5}{6}$.

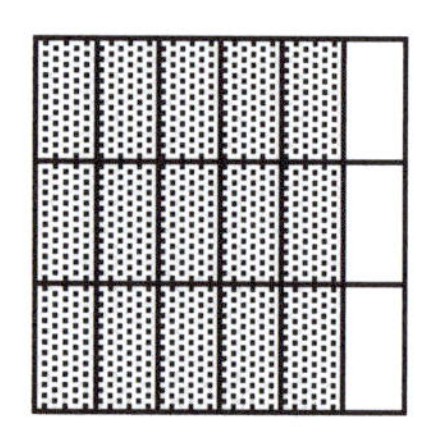

Step 2 Divide into 3 equal parts.

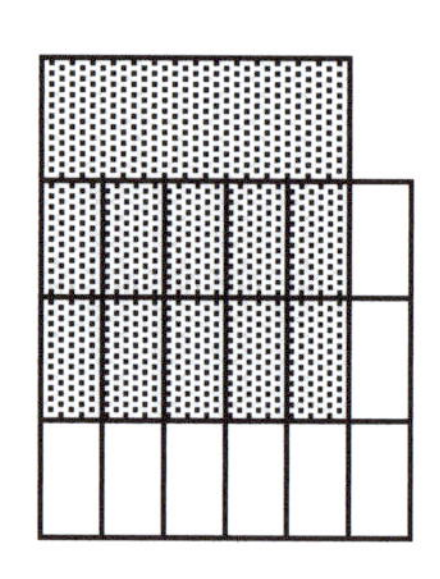

Step 3 Count 2 of those parts.

Pull up the violet insert to do this.

So $\frac{2}{3}$ of $\frac{5}{6}$ is $\frac{10}{18}$.

Another way to understand the formula for a fraction of a fraction (multiplying fractions) is shown in figure 1, which reveals the finished answer to example 2. To make the "factors" clearer I placed a white piece over the answer so you can see just the two factors of the two rectangles. I hope this helps you "see" numerator times numerator, divided by denominator times denominator.

Figure 1

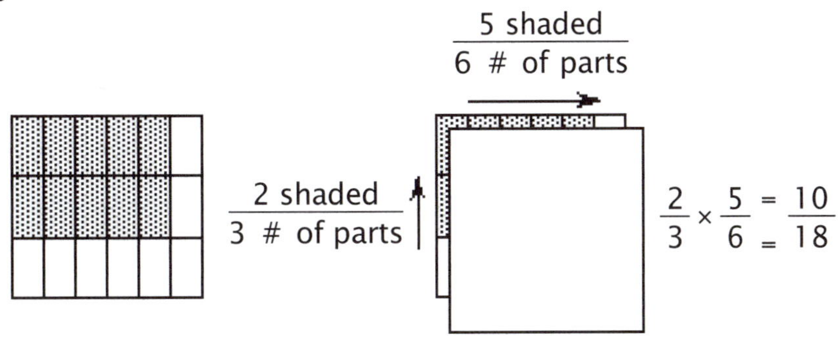

The whole square is an illustration of 2/3 x 5/6, which is a multiplication problem. In this problem we can see the two smaller problems. The numerator problem has the factors 2 x 5. The denominator problem has the factors 3 x 6. This exercise reveals the formula for multiplying two fractions.

$$\frac{2}{3} \times \frac{5}{6} = \frac{2 \times 5}{3 \times 6} = \frac{10}{18}$$

$$\frac{\text{Numberator times numerator (shaded rec tangle)}}{\text{Deno min ator times deno min ator (total \# of parts)}}$$

MENTAL MATH

Here are some more mental math problems for you to read aloud to your student. Try a few at a time, going slowly at first.

1. Four plus five, times two, divided by three, equals ? (6)
2. Thirty-six divided by four, minus two, times five, equals ? (35)
3. Sixty-six divided by six, plus four, divided by three, equals ? (5)
4. Forty-eight divided by eight, times six, plus two, equals ? (38)
5. Six plus seven, minus four, times nine, equals ? (81)
6. Nineteen minus three, divided by four, times seven, equals ? (28)
7. Nine times five, plus three, divided by six, equals ? (8)
8. Twenty-one divided by seven, plus one, times five, equals ? (20)
9. Three times four, divided by two, times nine, equals ? (54)
10. Seven times six, plus two, divided by 11, equals ? (4)

LESSON 10

Dividing Fractions with the Rule of Four

I will now introduce what some say is a radical thought. When dividing fractions, find the common denominator and divide the numerator. The concept of inverting the second fraction and multiplying, which is usually presented, is very hard to understand and explain to your student(s). The new method, while not as quick as inverting and multiplying, is much easier to understand and more in keeping with what a student typically knows at this stage. Not that we can't or shouldn't divide fractions the shortcut way. Inverting and multiplying, or multiplying by the reciprocal, is okay for experienced students, but not for beginners. Wait until they understand what you are doing before teaching this method. It will be presented in this book in lesson 23.

What we are trying to do is teach the concept and increase our understanding of why we do what we do. Students now know that to add or subtract fractions they must have a same, or common, denominator, then add or subtract the numerators. We will see that this strategy continues to work for division of fractions.

How you verbalize this concept is very important. When we learned $6 \div 2$ we said, "How many twos can we count out of six?" The answer is three. The same question works for dividing fractions. Try $4/5 \div 1/5$. Take one-fifth in one hand and four-fifths in the other hand, then ask the question, "How many one-fifths can I count out of four-fifths?" You can clearly see that the answer is four. (See example 1.) Then try $3/4 \div 1/4$. This is verbalized as, "How many one-fourths can I can count out of three-fourths?" The answer is three. (See example 2.)

How many quarters are there in one dollar? The equation is $1 \div 1/4$. We can't figure this until we give the parts the same denominator. Changing 1 to 4/4, we say, "How many one-fourths can we count out of four-fourths?" The answer is four. So the formula that is emerging is this: get the same denominator and divide the numerators. Notice that the denominators, once they are the same, are not important. 4/5 ÷1/5 is the whole number four.

Example 1

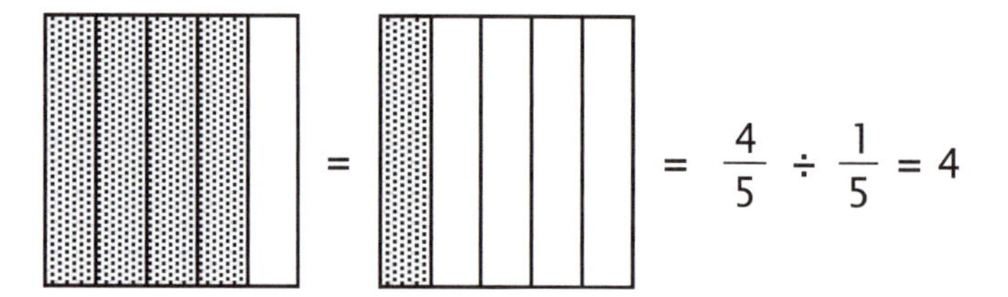

$$= \frac{4}{5} \div \frac{1}{5} = 4$$

Example 2

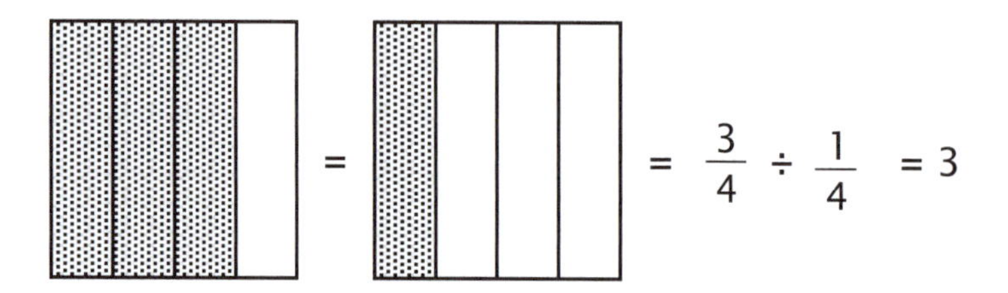

$$= \frac{3}{4} \div \frac{1}{4} = 3$$

To help students remember that the denominators are not important once they are the same, I've suggested they divide the numerator and denominator simultaneously. They will soon see that the denominator is always one with this method. Remember that anything divided by one is the number itself and does not affect the answer. Observe this principle in example 3.

Example 3

$$\frac{4 \div 1}{5 \div 5} = \frac{4}{1} = 4$$

Word Problems - Have the student carefully observe the language used in the division word problems on the lesson practice pages. Rather than looking for clue words, have the student read carefully and try to visualize what is being asked in each question. Systematic review pages 10D and 10E have specially marked word problems to illustrate the four basic operations using fractions.

Let's do a difficult problem like 4/5 ÷ 1/3. Using the overlays and our question when doing division, the problem appears like this:

Example 4

Solve 4/5 divided by 1/3.

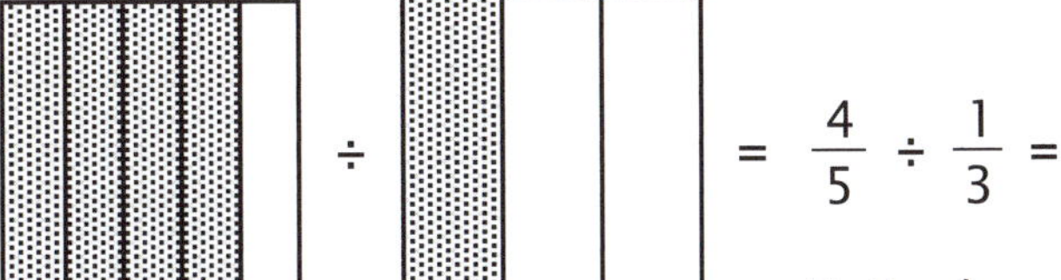

$$= \frac{4}{5} \div \frac{1}{3} =$$

Notice how difficult it is to estimate in this form.

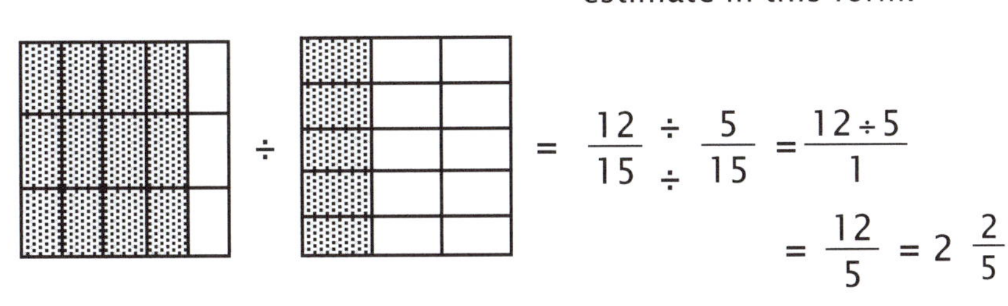

$$= \frac{12 \div 5}{15 \div 15} = \frac{12 \div 5}{1}$$

$$= \frac{12}{5} = 2\frac{2}{5}$$

12/5 is the same as 12 ÷ 5 or 2 2/5. You may write your answer in either form.

The question is, "How many one-thirds (or five-fifteenths) can I count out of four-fifths (or twelve-fifteenths)?" Look at the picture and ask, "How many groups of five can I count out of 12?" The result is two groups of five, and two of the fifths remaining, or 2 2/5.

Example 5

Solve $\frac{2}{3}$ divided by $\frac{1}{4}$.

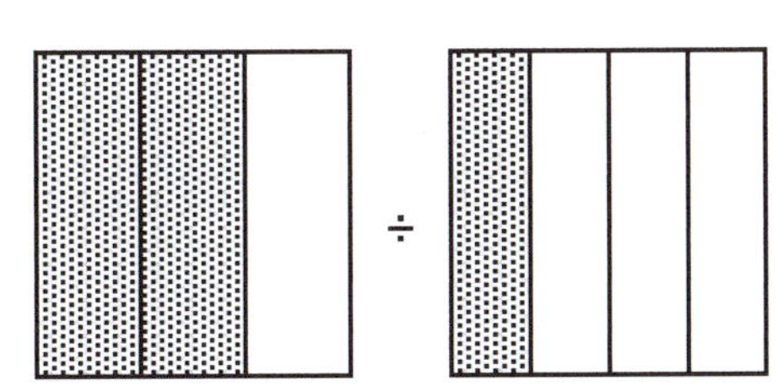

$$\frac{2 \div 1}{3 \div 4} =$$

Notice how difficult it is to estimate in this form.

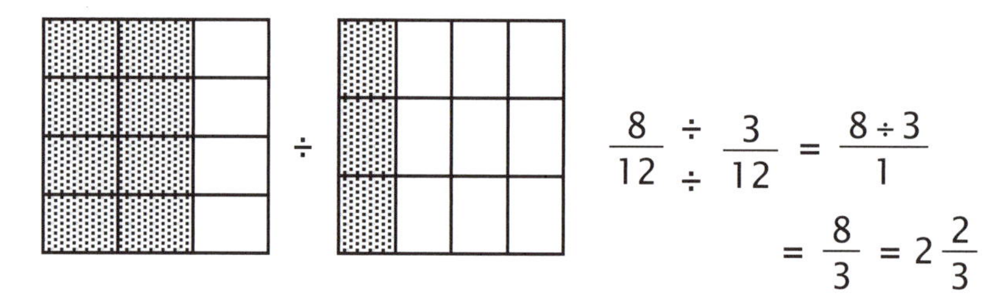

$$\frac{8}{12} \begin{matrix} \div \\ \div \end{matrix} \frac{3}{12} = \frac{8 \div 3}{1}$$

$$= \frac{8}{3} = 2\frac{2}{3}$$

8/3 is the same as 8 ÷ 3 or 2 2/3. You may write your answer in either form.

Look at the picture and ask, "How many groups of three can I count out of eight?"

Example 6

Solve $\frac{1}{4}$ divided by $\frac{3}{5}$.

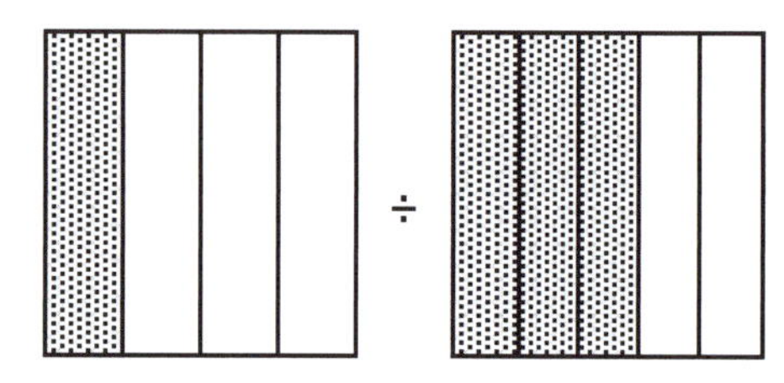

$$\frac{1}{4} \begin{matrix} \div \\ \div \end{matrix} \frac{3}{5} =$$

The question is, "How many three-fifths are there in one-fourth?"

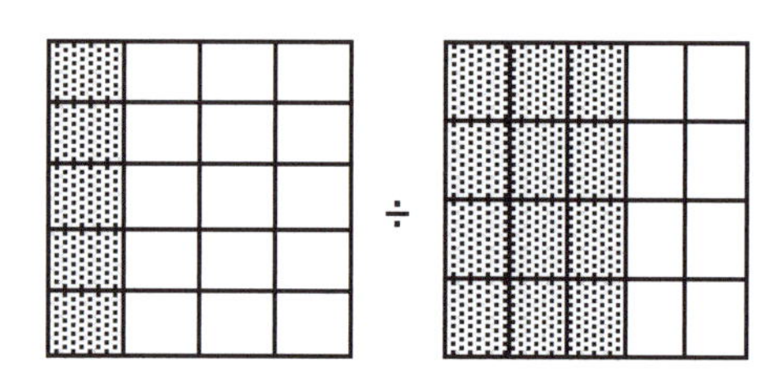

$$\frac{5}{20} \begin{matrix} \div \\ \div \end{matrix} \frac{12}{20} = \frac{5 \div 12}{1} = \frac{5}{12}$$

Remember that a fraction is another way of writing a division problem. In this case we are dividing a smaller number by a larger, so the answer is less than one.

LESSON 11

Finding Common Factors and Divisibility

Factoring is the opposite of multiplying. To multiply you are given two factors and you build a rectangle to find the area. To factor you are given the area, and you build a rectangle and find the factors.

Example 1

Find the factors of 6.

1 or 2

6 3

The factors are 1 x 6 and 2 x 3.

Example 2

Find the factors of 12.

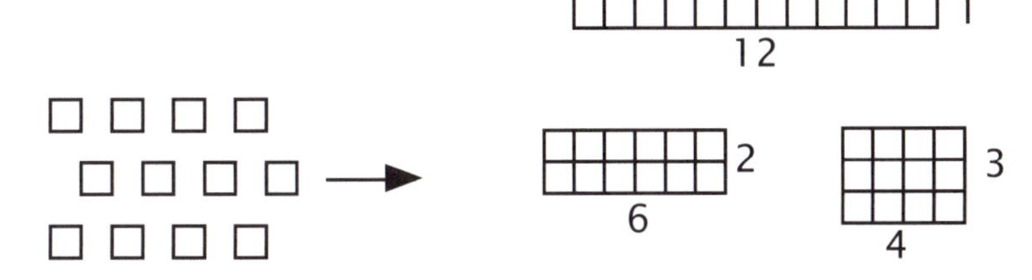

The factors are 1 x 12, 2 x 6, and 3 x 4.

Divisibility - Look at the numbers 18 and 24. What do you notice about them? You can tell a lot about a number by looking at the number in the units place. If a number ends in 2, 4, 6, 8, or 0, it is an even number. Even numbers are always divisible by two. "Divisible by two" means that you can divide the number by two with no remainder. It also means that at least one of its factors is two.

2	12
4	14
6	16
8	18
10	20

If a number ends in zero, it is a multiple of 10. A *multiple* is a result of multiplying. If a number is a multiple of 10, it means that at least one of its factors is 10. Patterns emerge when you look at the skip counting facts. Look at the 10 facts below and notice that each number ends in zero and is a multiple of 10.

10	60
20	70
30	80
40	90
50	100

When studying the five facts we can see that all the facts have a five or a zero in the units place. So we conclude that if a number ends in five or zero it is divisible by five, or at least one of its factors is five.

5	30
10	35
15	40
20	45
25	50

Twos, fives, and tens are easily seen by looking at the number in the units place. To learn the pattern in the threes and nines, a different approach is needed. Looking at the three facts on the next page, notice that if you add up the digits they add up to three or a multiple of three. To know whether a number is divisible by three, simply add up the digits, even for very large numbers. If you add up the digits and they add up to three or a multiple of three, then the number is divisible by three.

3		18	1+8= 9
6		21	2+1= 3
9		24	2+4= 6
12	1+2= 3	27	2+7= 9
15	1+5= 6	30	3+0= 3

Example 3

Is 4,715 a multiple of 3?

4 + 7 + 1 + 5 = 17.

17 is not a multiple of 3, so 4,715 is not a multiple of 3.

Example 4

Is 1,218 a multiple of 3?

1 + 2 + 1 + 8 = 12.

12 is a multiple of 3, so 1,218 is a multiple of 3.

To know whether a number is divisible by nine simply add up the digits. If the digits add up to nine or a multiple of nine, then it is divisible by nine.

9	0 + 9 = 9	54	5 + 4 = 9
18	1 + 8 = 9	63	6 + 3 = 9
27	2 + 7 = 9	72	7 + 2 = 9
36	3 + 6 = 9	81	8 + 1 = 9
45	4 + 5 = 9	90	9 + 0 = 9

Example 5

Is 7,218 a multiple of 9?

7 + 2 + 1 + 8 = 18.

18 is a multiple of 9, so 7,218 is a multiple of 9.

Example 6

Is 8,729 a multiple of 9?

8 + 7 + 2 + 9 = 26.

26 is not a multiple of 9, so 8,729 is not a multiple of 9.

There are other, more complex methods for testing if a number is divisible by four, six, or eight, but these methods add needless work to the process. If a number is a multiple of four, you can divide by two twice. If a number is a multiple of six, then it must also be a multiple of two and three. So divide by two, then by three.

If you learn how to recognize the multiples of 2, 3, 5, 9, and 10, this knowledge will serve you well when reducing fractions.

There are three words to consider in "*greatest common factor*": the *factor*, the *common* factor, and the *greatest* common factor. The factors of six are 1 x 6 and 2 x 3. These can be written in order as 1, 2, 3, and 6. The common factors of six and 15 are one and three. They are underlined in figure 1.

Figure 1

6	1, 2, 3, 6
15	1, 3, 5, 15

The *greatest* common factor (GCF) or largest of the common factors is three.

Example 1

Find the GCF of 12 and 18.

1. The factors of 12 and 18 are as follows:

12	1, 2, 3, 4, 6, 12
18	1, 2, 3, 6, 9, 18

2. The common factors of 12 and 18 are underlined.

12	1, 2, 3, 4, 6, 12
18	1, 2, 3, 6, 9, 18

3. The factors 1, 2, 3, and 6 are each common factors of 12 and 18, but 6 is the greatest common factor.

Example 2

Find the GCF of 18 and 27.

1. The factors of 18 and 27 are as follows:

   18 1, 2, 3, 6, 9, 18
   27 1, 3, 9, 27

2. The common factors of 18 and 27 are underlined.

   18 1, 2, 3, 6, 9, 18
   27 1, 3, 9, 27

3. The factors 1, 3, and 9 are each common factors of 18 and 27, and 9 is the greatest common factor.

Example 3

Find the GCF of 30 and 45.

1. The factors of 30 and 45 are as follows:

   30 1, 2, 3, 5, 6, 10, 15, 30
   45 1, 3, 5, 9, 15, 45

2. The common factors of 30 and 45 are underlined.

   30 1, 2, 3, 5, 6, 10, 15, 30
   45 1, 3, 5, 9, 15, 45

3. The factors 1, 3, 5, and 15 are each common factors of 30 and 45, and 15 is the greatest common factor.

LESSON 12

Reducing Fractions—Common Factors

Word Problems

Reducing fractions is the inverse of making equivalent fractions. The two words I use to indicate this inverse relationship are *increasing* and *decreasing*. When making equivalent fractions, we make a fraction, for example, one-half, then place overlays horizontally on top to change the number of pieces without affecting the amount. Same amount, more pieces—remember? What we want to focus on is what happens when we place this overlay on top, or how we record this. Look at the picture below and the corresponding equations.

Example 1

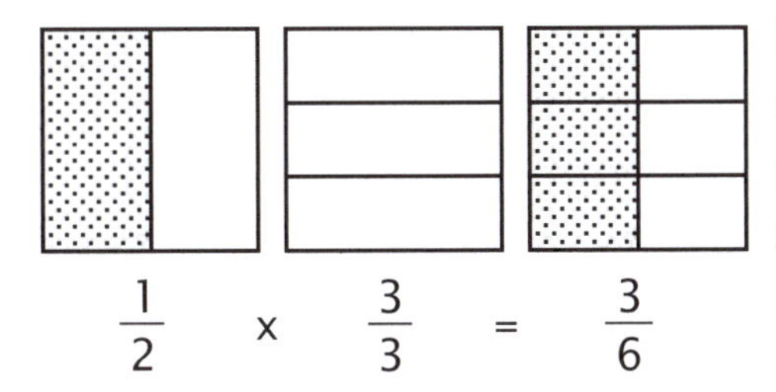

Beginning with 1/2, the numerator and denominator are each increased by a common factor of three.

In example 1 the overlay is placed on top and increases the numerator by a factor of three and the denominator by a factor of three. Notice that even though you are placing only one overlay on top, both the numerator and denominator are being increased by the same common factor, in this case three.

Another way to look at this is to notice the over factor and the up factor. In example 2, the picture from example 1 is covered to reveal the edges or factors. It is over 1/2 and up 3/3. Recognize that 3/3 is equivalent to one. It is the same amount but more pieces, because you are multiplying by one. Any value multiplied by one is still the same value. It has more pieces, but it is the same amount. Be sure to emphasize this.

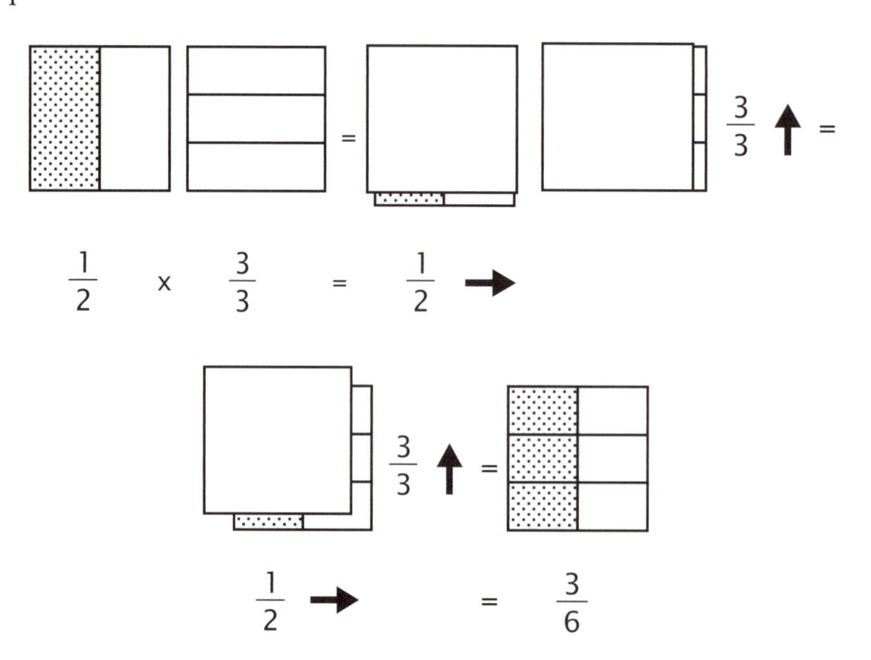

Example 2

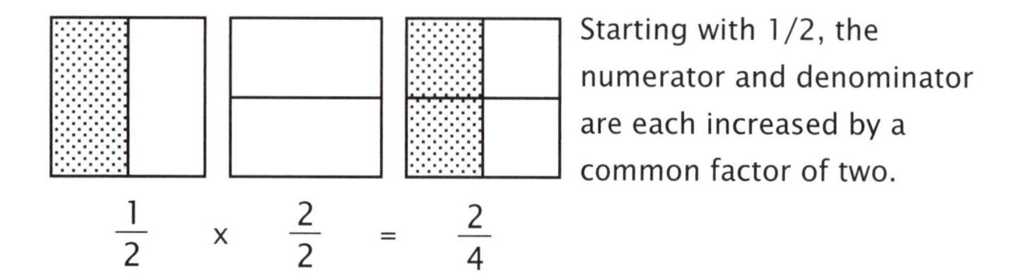

Starting with 1/2, the numerator and denominator are each increased by a common factor of two.

Example 3

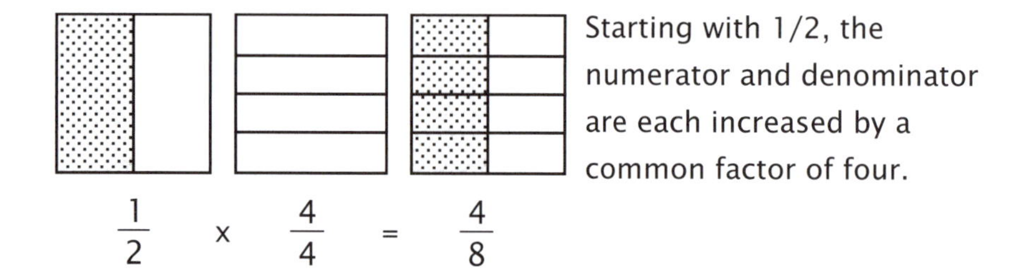

Starting with 1/2, the numerator and denominator are each increased by a common factor of four.

If we reverse this process, we divide the numerator and the denominator by the same factor, which is decreasing, or reducing. The pictures on the next page show reducing, or decreasing, by a common factor.

Example 4

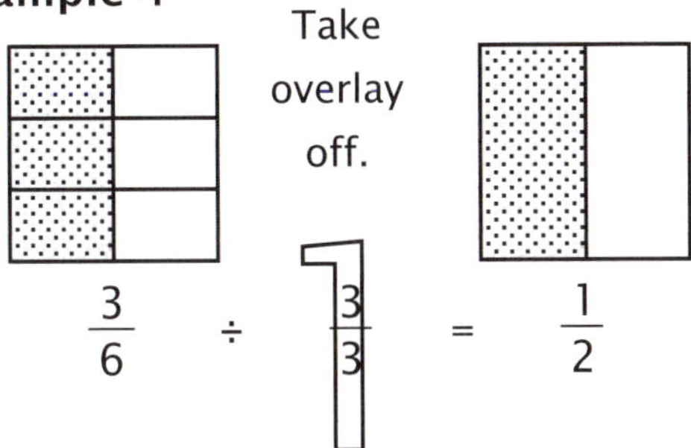

3/6 is divided by one, which means the numerator and denominator are reduced by a common factor of three.

Example 5

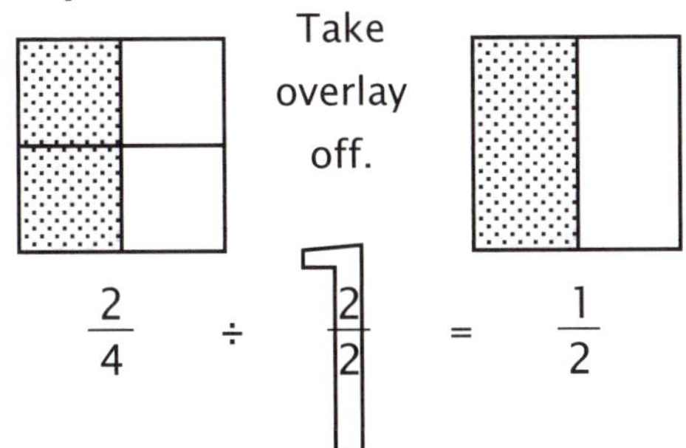

2/4 is divided by one, which means the numerator and denominator are reduced by a common factor of two.

Example 6

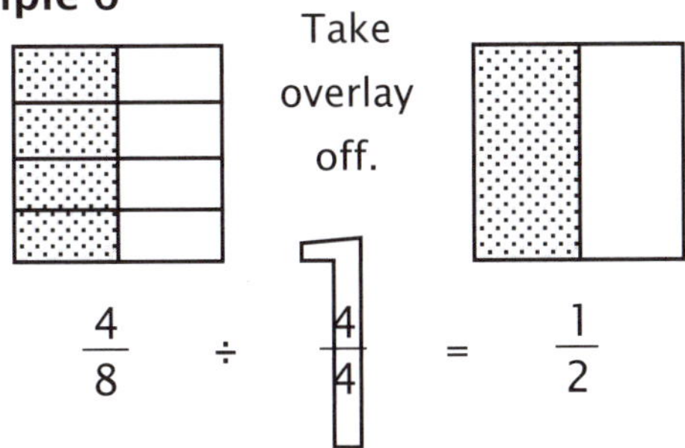

4/8 is divided by one, which means the numerator and denominator are reduced by a common factor of four.

Remembering what we observed in example 2 on the previous page, it is the up factor that either increases or reduces a fraction. Looking at the up factor tells us by what common factor the fraction is being increased, or by what common factor it is being reduced.

WORD PROBLEMS

Here are a few more multistep word problems to try. You may want to read and discuss these with your student as you work out the solutions together. Again, the purpose is to stretch, not to frustrate. If you do not think the student is ready, you may want to come back to these later.

The answers are at the end of the solutions section at the back of this book.

1. Jennifer had $30 to spend on herself. She spent one-fifth of the money on a sandwich, one-sixth for a ticket to a museum, and one-half of it on a book. How much money does Jennifer have left over?

2. Mark drove for one-half of the trip, and Justin drove for one-fourth of the trip. Gina and Kaitlyn divided the rest of the driving evenly between them. If the entire trip was 128 miles, how many miles did Kaitlyn drive?

3. Two pie pans of the same size remain on the counter. One pan has one-half of a pie, and the other has three-fourths of a pie left in it. Mom wishes to divide all the pie into pieces that are each one-eighth of a pie. How many pieces of pie will she have when she is finished?

LESSON 13

Reducing Fractions—Prime Factors

If you are factoring and can build only one rectangle, as with five or seven, you have a *prime number*. A number from which you can build more than one rectangle, such as six or 12, is a *composite number*. The composite numbers between one and 24 are 4, 6, 8, 9, 10, 12, 14, 15, 16, 18, 20, 21, 22, and 24.

The definition of a prime number is "any number that has only the factors of one and itself, or that can only be divided evenly by one and itself."

To illustrate this concept, give the students seven green blocks and offer a reward for the first person who can make two different rectangles using all of the blocks in each rectangle. They will soon find out that only one rectangle can be built. Note that a rectangle that is one by seven is the same as a rectangle that is seven by one. One is standing and the other is reclining. I then say, "What do we call a number that you can only build one rectangle with?" The answer is a prime number. Then I recite the definition about a number having only factors (or dimensions) of one and itself, or only being able to be divided evenly by one and itself. Having seen this relationship with the manipulatives, the definition now makes sense.

Example 1

Find all the possible factors of seven and tell whether the number is prime or composite.

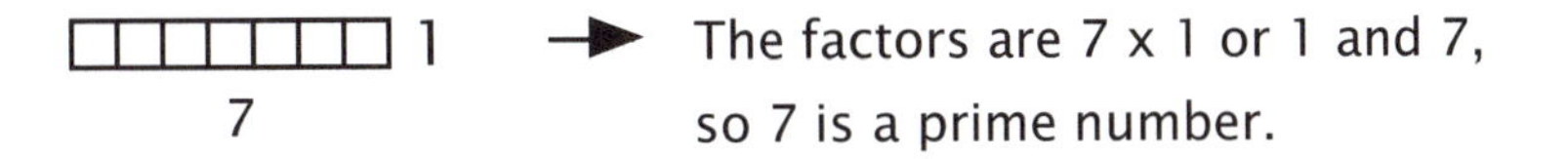

The prime numbers between one and 24 are 2, 3, 5, 7, 11, 13, 17, 19, and 23. Choose any of these to ask the students to build, and they will find that there is only one rectangle to be built, and thus only one set of factors.

Note that one is not considered a prime number.

The prime factors of a number may be discovered *(prime factorization)* using a factor tree or repeated division. Here is an example of each as we find the prime factors of 12 and 18.

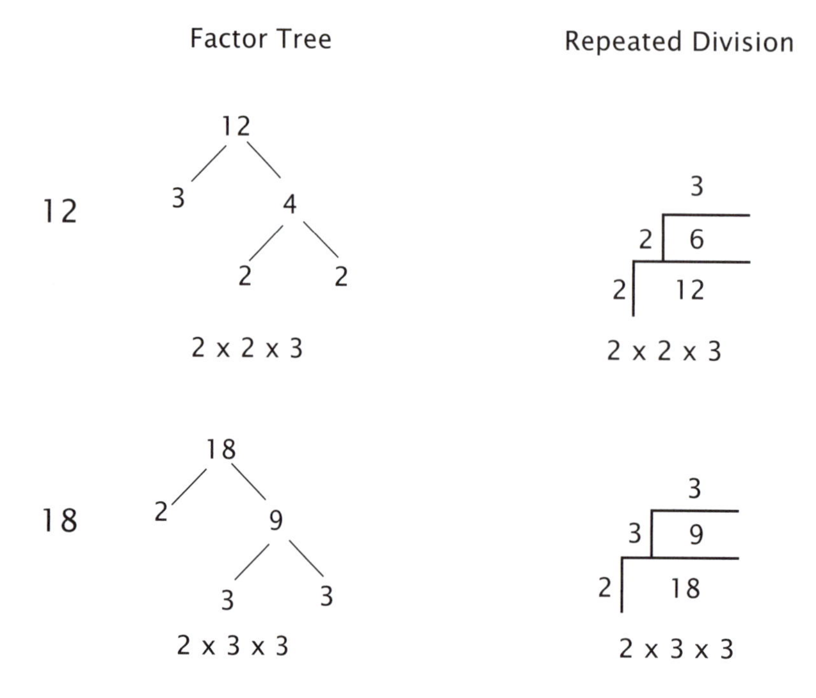

Example 2

Reduce $\frac{12}{18}$ using the prime factorization method.

We know the prime factors of 12 and 18 from the factor tree and repeated division. Now we have to find which numbers in the numerator and denominator may be divided to make one.

$$\frac{12}{18} = \frac{2 \times 2 \times 3}{2 \times 3 \times 3} = \frac{2}{2} \times \frac{2}{3} \times \frac{3}{3} = 1 \times 1 \times \frac{2}{3}$$

Notice the outline of one behind 2/2 and 3/3. The fraction 2/2 (two divided by two) equals one and 3/3 equals one. Some refer to the twos and threes disappearing as "canceling." They don't really cancel or disappear, so I prefer to say that 2/2 and 3/3, being one, are absorbed into 2/3.

What makes this method of reducing attractive is not having to recognize the greatest common factor. In some cases the GCF is very obvious, and then you should use it to reduce. When the greatest common factor isn't so obvious, simply find the prime factors and start dividing until the fraction is in its lowest terms and no more "ones" may be divided out of it.

Example 3

Reduce 10/14 using the prime factorization method.

First find all the prime factors of 10 and 14.

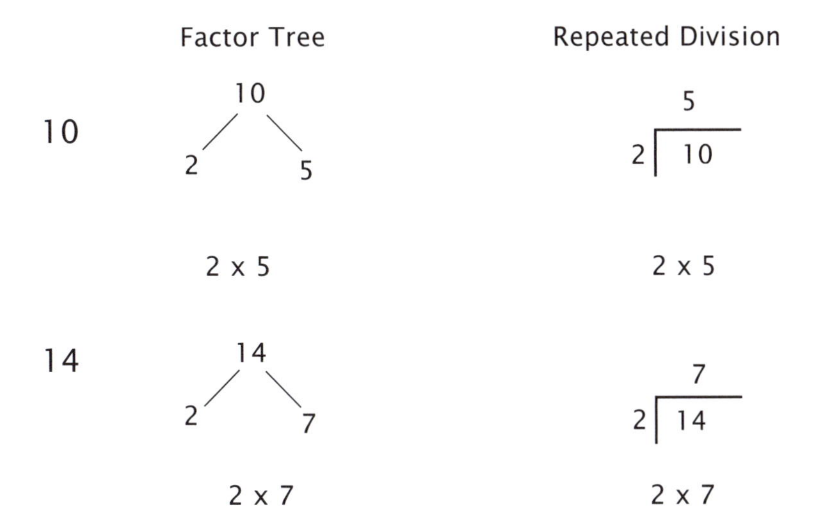

Then find which numbers in the numerator and denominator may be divided to make one.

$$\frac{10}{14} = \frac{2 \times 5}{2 \times 7} = \frac{2}{2} \times \frac{5}{7} = 1 \times \frac{5}{7}$$

This problem can be verbalized as two divided by two, times five divided by seven; and two divided by two is one.

LESSON 14

Linear Measure

When discussing three-fourths in a previous lesson, recall how we calculated a fraction of a number and a fraction of one. First we looked at what we were starting with, then divided it into equal parts (denominator), then counted the number of equal parts (numerator). In this lesson, we are taking 3/4 of one inch. We start with one inch, divide it into four equal parts, and count three of them.

This method of counting the total number of spaces within one inch to find the denominator, then counting a certain number of those spaces for the numerator, is exactly how fractions have been taught throughout this book. It will especially benefit the student when encountering rulers with sixteenths, fifths, and tenths. Instead of merely memorizing the halves, fourths, and eighths, the student will be able to read any ruler.

Here are a few "abnormal" measurement problems to make sure the concept is mastered.

When working through examples 1 and 2, use the overlays to create the fraction, pretending the width of the overlays represents one inch instead of five inches.

Example 1

Draw a line 4/5 of an inch, or 4/5".

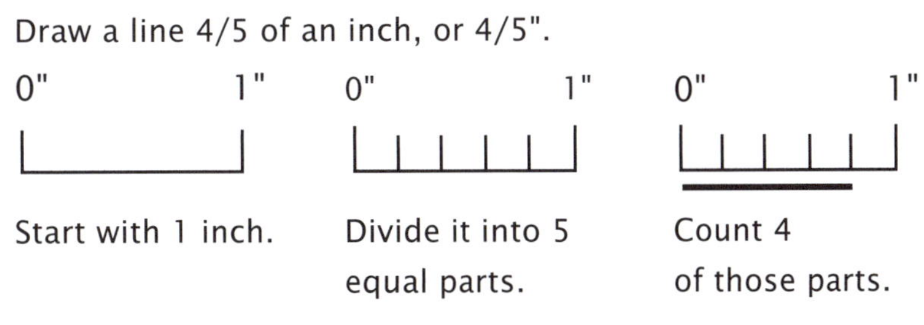

Example 2

Draw a line 2/3 of an inch, or 2/3".

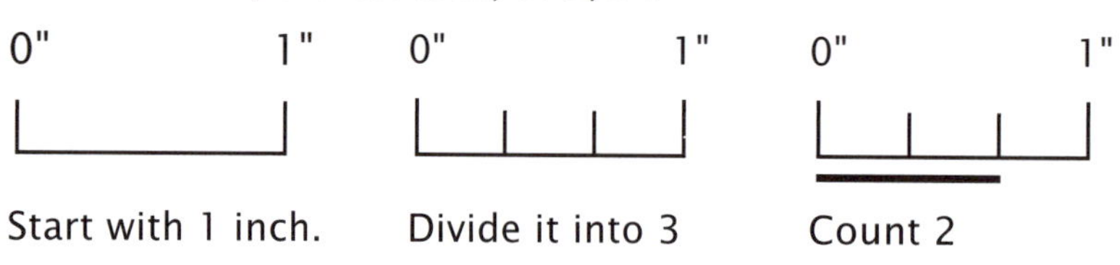

Start with 1 inch. Divide it into 3 equal parts. Count 2 of those parts.

Example 3

How long is the line segment?

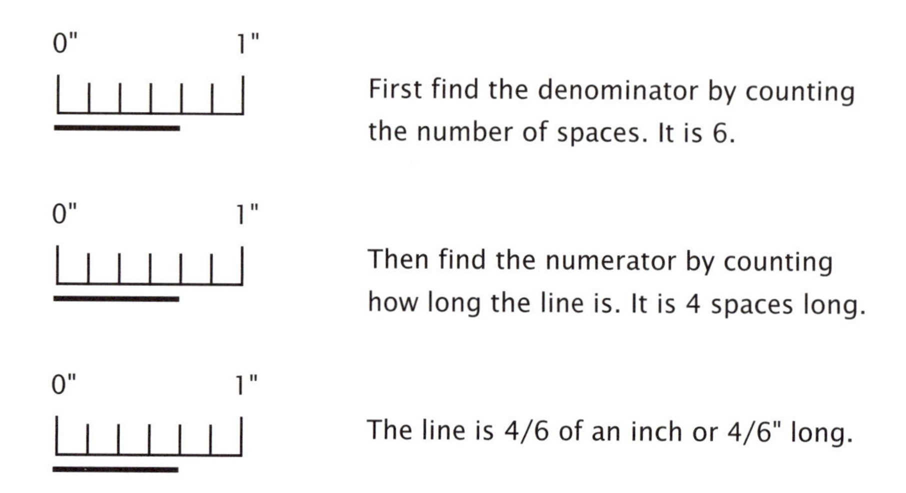

First find the denominator by counting the number of spaces. It is 6.

Then find the numerator by counting how long the line is. It is 4 spaces long.

The line is 4/6 of an inch or 4/6" long.

Example 4

How long is the line segment?

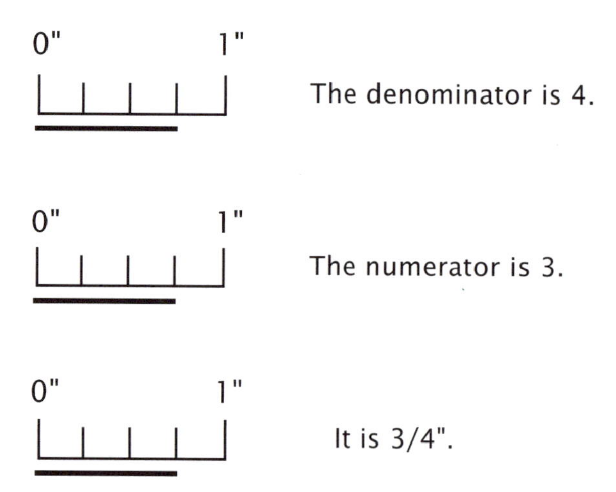

The denominator is 4.

The numerator is 3.

It is 3/4".

There are two 1/2" x 5" clear overlays to create eighths and sixteenths, which are used extensively in measurement. Place the 1/8 overlay on top of 1/2 to show that 1/2 equals 4/8. (See figure 1.) Place the 1/16 overlay on top of 1/2 to show that 1/2 is the same as 8/16. (See figure 2.) To show that 1/2 is equal to 2/4, use the 1/4 overlay. (See figure 3.)

Figure 1

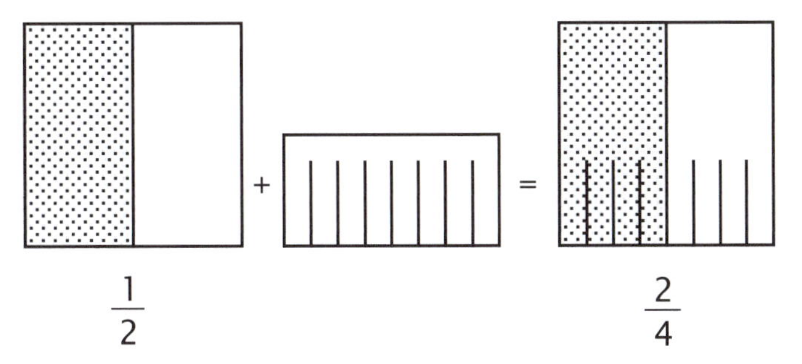

Figure 2

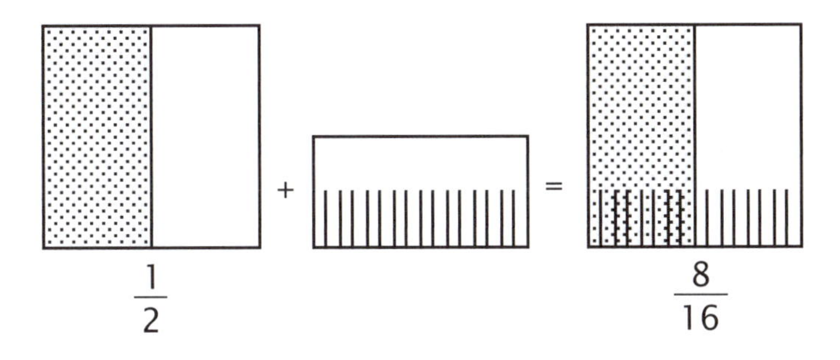

Figure 3

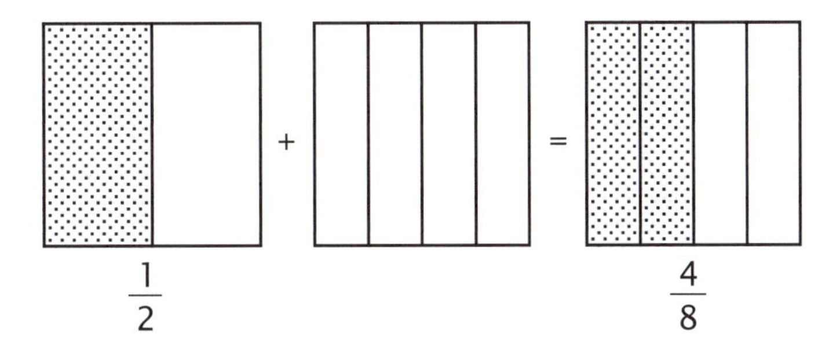

Example 5

Show that 3/4 = 6/8 = 12/16.

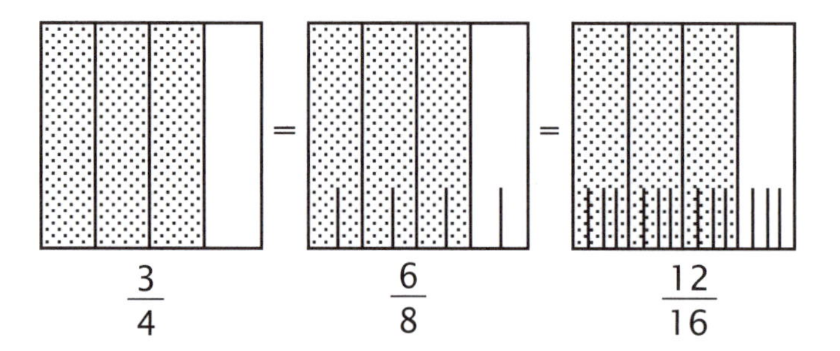

Having made eighths and sixteenths, notice how reducing fractions applies to measurement. The measurements 8/16", 4/8", and 2/4" may all be expressed as 1/2". This concept is used with open-end wrenches. There is a 1/2" wrench, a 5/8" wrench, and a 3/4" wrench, to name a few. Notice that there aren't 2/8", 4/8", and 6/8" wrenches or 8/16", 10/16", and 12/16" wrenches. Always reduce when measuring.

A succession of rulers follows. Observe how the last ruler is a compilation of the first four rulers. There are different types of rulers, but this one is the most common in the U.S.

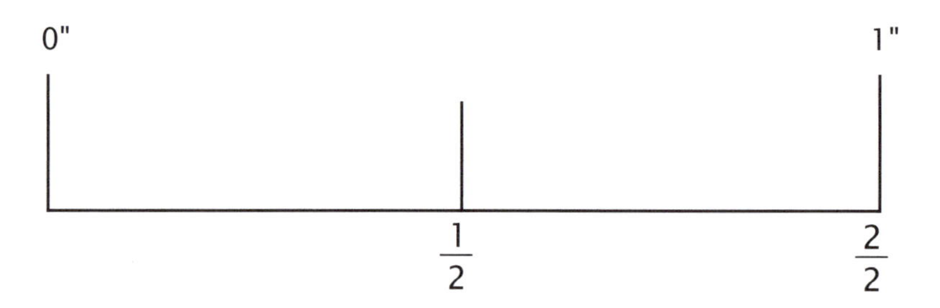

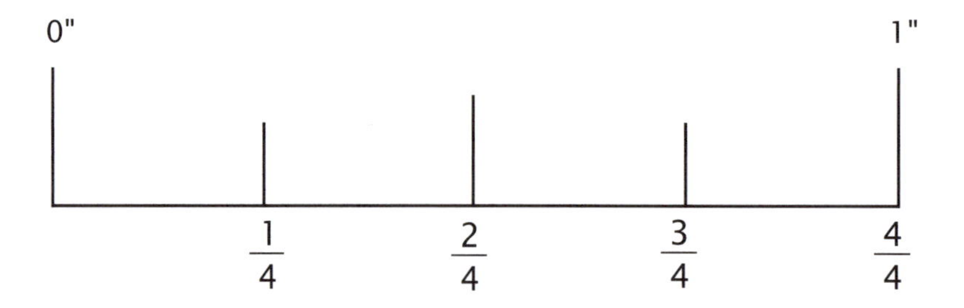

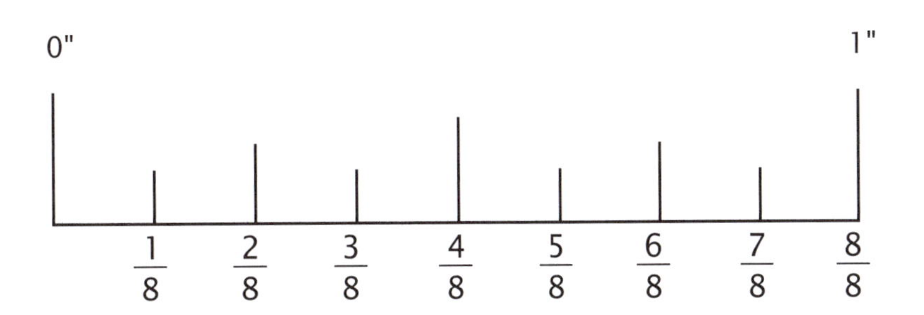
0"
1"
1/8
2/8
3/8
4/8
5/8
6/8
7/8
8/8

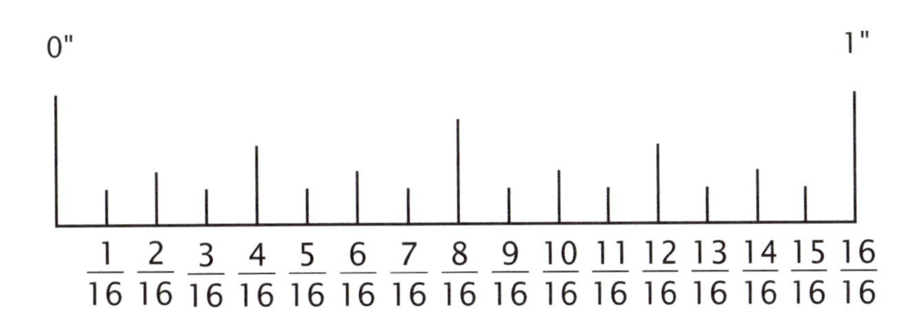
0"
1"
1/16
2/16
3/16
4/16
5/16
6/16
7/16
8/16
9/16
10/16
11/16
12/16
13/16
14/16
15/16
16/16

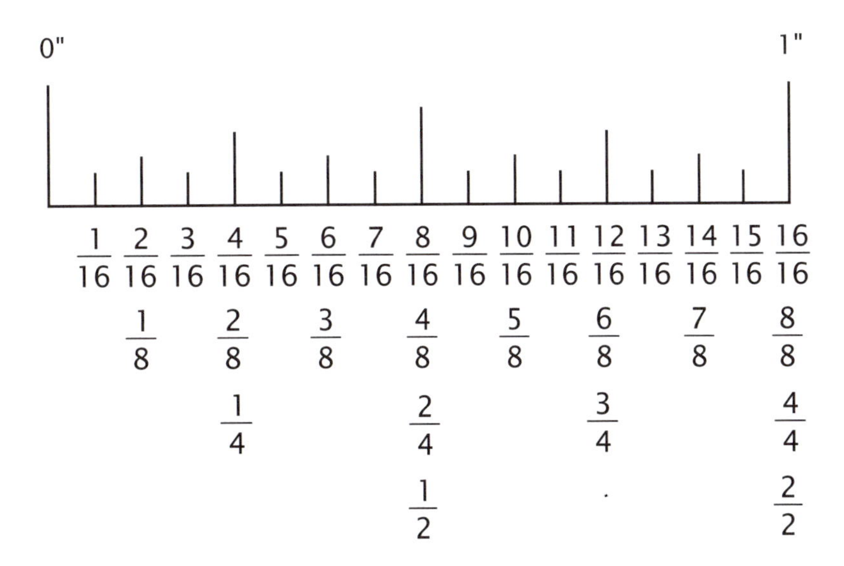
0"
1"
1/16
2/16
3/16
4/16
5/16
6/16
7/16
8/16
9/16
10/16
11/16
12/16
13/16
14/16
15/16
16/16
1/8
2/8
3/8
4/8
5/8
6/8
7/8
8/8
1/4
2/4
3/4
4/4
1/2
2/2

LESSON 15

Mixed Numbers and Improper Fractions

Mental Math

A *mixed number* is a combination of a whole number and a fraction. Maybe mixed numbers should be called frac-bers (*frac* from fraction and *ber* from number), or num-tions (*num* from number and *tion* from fraction). $2\frac{1}{2}$ is a mixed number: 2 is the number and $\frac{1}{2}$ is the fraction.

If a fraction has a numerator smaller than the denominator, it is called a *proper fraction*. If the numerator is the same as the denominator, the fraction is equal to one. If the numerator is larger than the denominator, the fraction is an *improper fraction*. I mention this now because when you change a mixed number to a fraction, the resulting fraction will always be an improper fraction.

To change a mixed number to an improper fraction, use the green colored pieces, the white pieces, or the other colored pieces that are the same size to represent one. In the thirds pocket the largest pink piece is used to show 3/3, or one. In the fifths pouch, the largest blue piece represents 5/5, or one.

Some students may be helped to understand changing a mixed number to an improper fraction by considering money. If you have two dollars and a quarter, how many quarters do you have? The answer is four quarters for each dollar, plus one quarter, or nine quarters altogether. See example 1. The expression 2 1/4 is a mixed number, and 9/4 is an improper fraction.

Example 1

$$2\frac{1}{4} = 1 + 1 + \frac{1}{4} = \frac{4}{4} + \frac{4}{4} + \frac{1}{4} = \frac{9}{4}$$

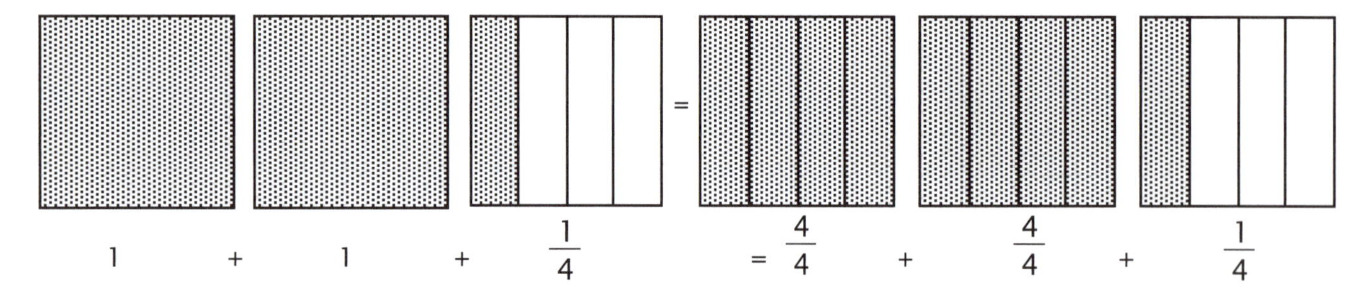

Example 2

Change 1 3/5 to an improper fraction.

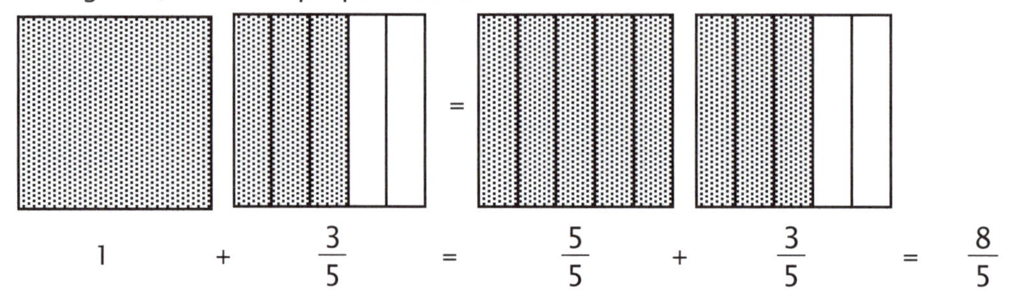

1 3/5 is a mixed number, and 8/5 is an improper fraction.

To change an improper fraction to a mixed number, do the opposite. I find that thinking about money and quarters is very helpful in understanding this process. Ask the student, "If you have nine quarters, how many dollars do you have?" There are four quarters in a dollar, so eight quarters is two dollars, and you have one quarter left over. Using the overlays it looks like this:

Example 3

Change 9/4 to a mixed number.

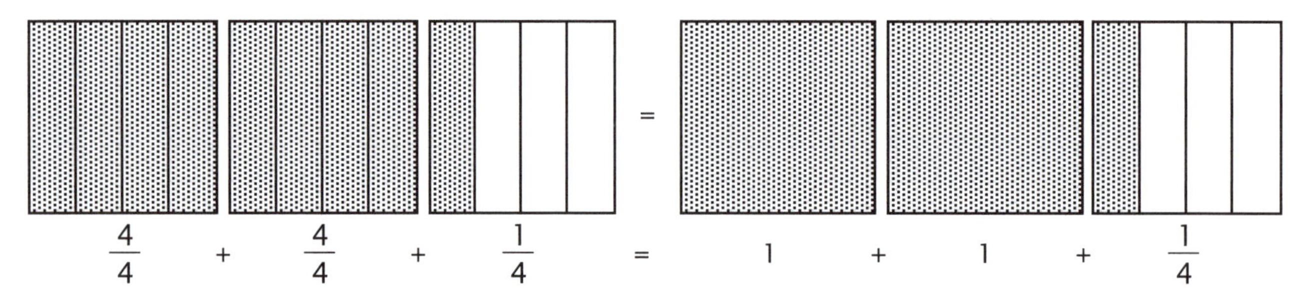

Start with 9/4 and subtract 4/4 (which is one) to get 5/4.

Then subtract 4/4 again (another one) and 1/4 is left.

So 9/4 is the same as 2 and 1/4.

Example 4

Change 12/5 to a mixed number.

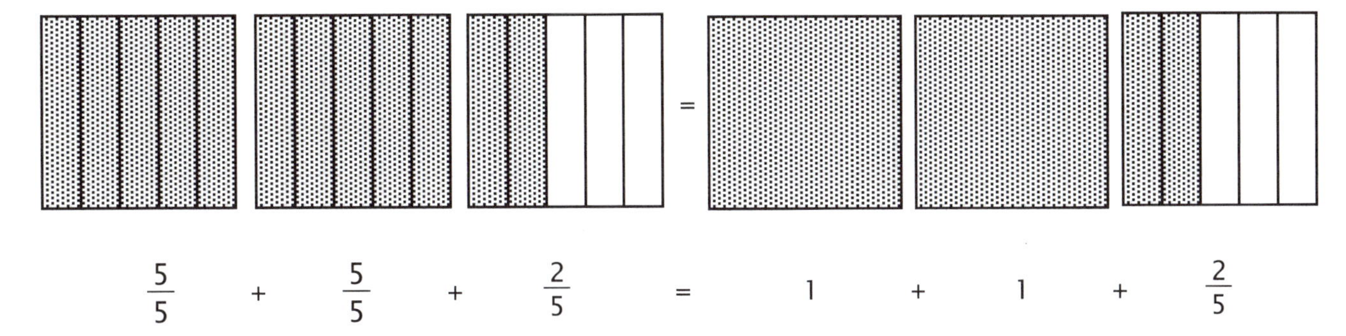

Start with 12/5 and subtract 5/5 (which is one) to get 7/5. Then subtract 5/5 again (another one) and 2/5 is left. So 12/5 is the same as 2 and 2/5.

Example 5

Change $\frac{5}{3}$ to a mixed number.

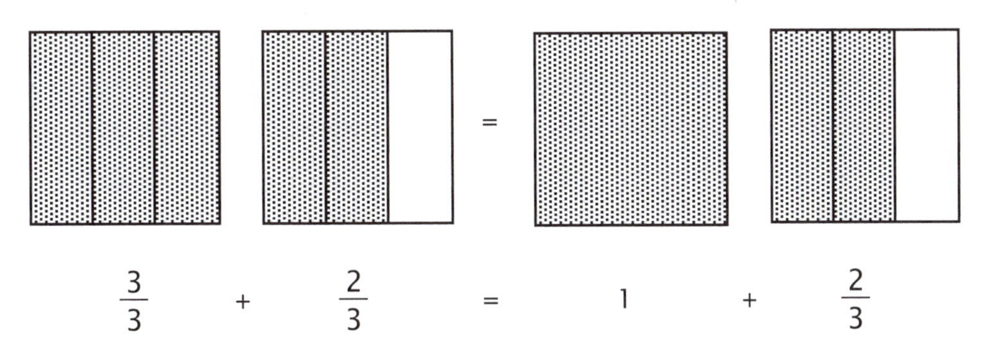

Start with 5/3 and subtract 3/3 (which is one) to get 2/3. So 5/3 is the same as one and 2/3.

MENTAL MATH

Here are some more mental math problems for you to read aloud to your student. Read the problems in the order given.

1. Twenty-four divided by four, plus two, times six, equals ? (48)
2. Fifteen minus eight, times two, plus five, equals ? (19)
3. Three plus three, times five, divided by 10, equals ? (3)
4. Twenty-five divided by five, times seven, plus four, equals ? (39)
5. Fifty minus one, divided by seven, plus two, equals ? (9)
6. Sixty-four divided by eight, minus two, times four, equals ? (24)
7. Twenty-five plus two, divided by three, times six, equals ? (54)
8. Three times five, plus five, divided by two, equals ? (10)
9. Fifty-six divided by seven, divided by two, times seven, equals ? (28)
10. Nine times nine, minus one, divided by 10, equals ? (8)

LESSON 16

Linear Measure with Mixed Numbers

In this unit we are combining what we now know about linear measure with what we just learned about mixed numbers. In linear measure we learned how to read a fraction of an inch. A mixed number employs a whole number along with a fraction of a number. To read a measuring tape, start at the left and read to the nearest whole number, in this case the nearest inch. Then read the fraction of the next inch. Putting the whole number of inches with the fraction of an inch produces the correct measure.

Example 1

How long is the line?

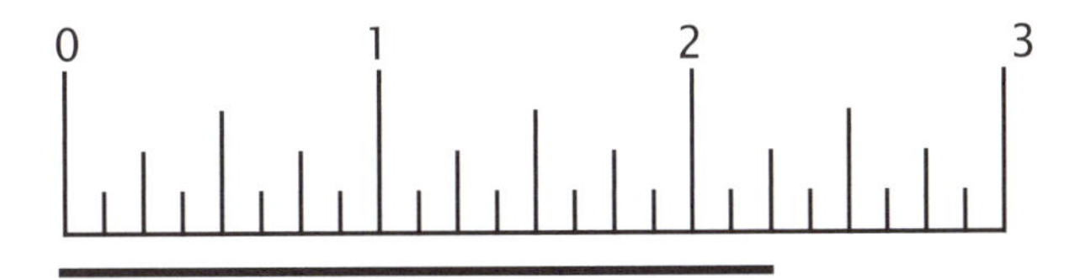

Starting at the left, we see two full inches.
Beginning at the two-inch mark, we read a fraction of an inch, which is 1/4".
Putting them together, we see the line is 2 1/4".

Example 2

How long is the line?

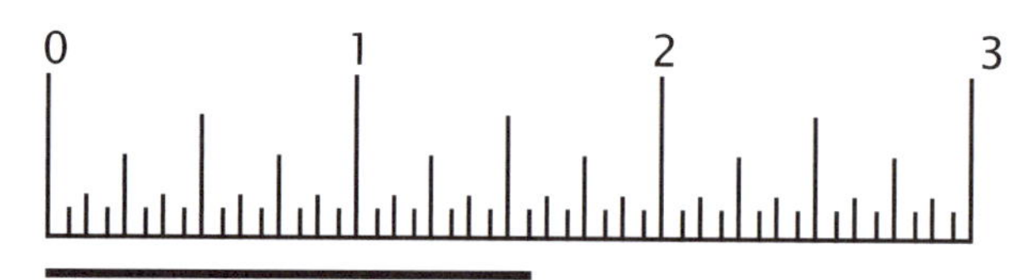

Starting at the left, we see one full inch.
Beginning at the one-inch mark, we read a
fraction of an inch, which is 9/16".
Putting them together, we see the line is 1 9/16".

Example 3

How long is the line?

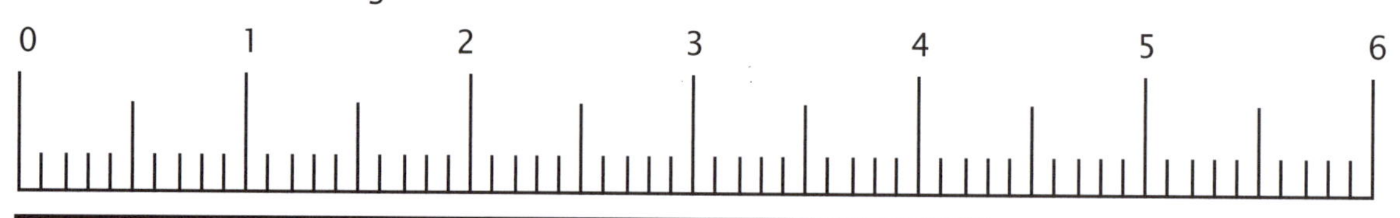

Starting at the left, we see four full inches.
Beginning at the four-inch mark, we read a
fraction of an inch, which is 3/10".
Putting them together, we see the line is 4 3/10".

Example 4

How long is the line?

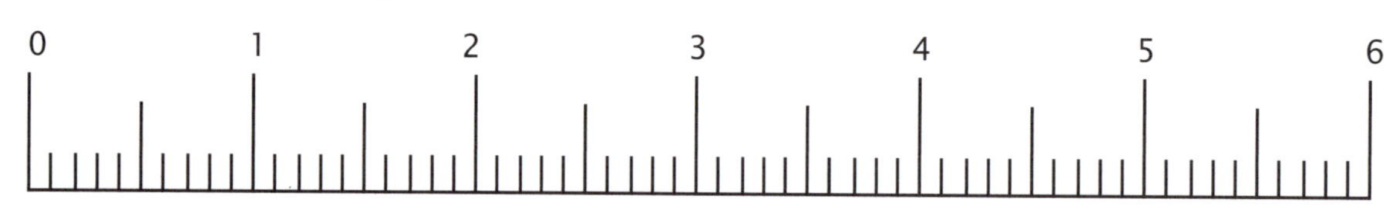

Starting at the left, we see five full inches.
Beginning at the five-inch mark, we read a
fraction of an inch, which is 5/10", which reduces to 1/2".
Putting them together, we see the line is 5 1/2".

LESSON 17

Addition and Subtraction of Mixed Numbers

In order to combine two items, they must be the same kind. Stress this concept! When working with mixed numbers, we can combine, or add, numbers with numbers. We can also add fractions to fractions, provided they have the same denominator. So add the fractions to the fractions and the numbers to the numbers.

When working with mixed numbers, it is important to estimate your answer before solving. In example 1, we can round three-fifths to one, so 2 3/5 is almost three. We can round one-fifth to zero, so 4 1/5 is close to four. Adding the estimates: 3 + 4 = 7. The answer to example 1 should be close to, or approximately, seven. The symbol "≈" means approximately. The estimated numbers, 3 and 4, are circled.

Example 1

$$\begin{array}{rr} & 2\frac{3}{5} \quad (3) \\ + & 4\frac{1}{5} \quad (4) \\ \hline & 6\frac{4}{5} \quad (\approx 7) \end{array}$$

2 + 4 = 6 (number plus number)

$\frac{3}{5} + \frac{1}{5} = \frac{4}{5}$ (fraction plus fraction)

Example 2

$$\begin{array}{rr} & 5\frac{4}{7} \quad (6) \\ + & 3\frac{2}{7} \quad (3) \\ \hline & 8\frac{6}{7} \quad (\approx 9) \end{array}$$

5 + 3 = 8 (number plus number)

$\frac{4}{7} + \frac{2}{7} = \frac{6}{7}$ (fraction plus fraction)

Example 3

$$\begin{array}{r} 8\ \frac{9}{11} \ ⑨ \\ -\ 2\ \frac{5}{11} \ ② \\ \hline 6\ \frac{4}{11} \ (\approx 7) \end{array}$$

8 - 2 = 6 (number minus number)

$\frac{9}{11} - \frac{5}{11} = \frac{4}{11}$ (fraction minus fraction)

Example 4

$$\begin{array}{r} 7\ \frac{10}{13} \ ⑧ \\ -\ 5\ \frac{9}{13} \ ⑥ \\ \hline 2\ \frac{1}{13} \ (\approx 2) \end{array}$$

7 - 5 = 2 (number minus number)

$\frac{10}{13} - \frac{9}{13} = \frac{1}{13}$ (fraction minus fraction)

LESSON 18

Addition of Mixed Numbers

with Regrouping; Word Problems

Adding mixed numbers with regrouping is just like regrouping with whole numbers. Instead of converting 10 units to one ten, we are changing fractions to a number, in this case 3/3 to one. Adding 2 + 1 is no problem, but adding 2/3 to 2/3 produces 4/3, which is an improper fraction. The improper fraction 4/3 is the same as 3/3 plus 1/3, and 3/3 is equal to one. So 4/3 is the same as 1 1/3, and 3 + 1 1/3 = 4 1/3.

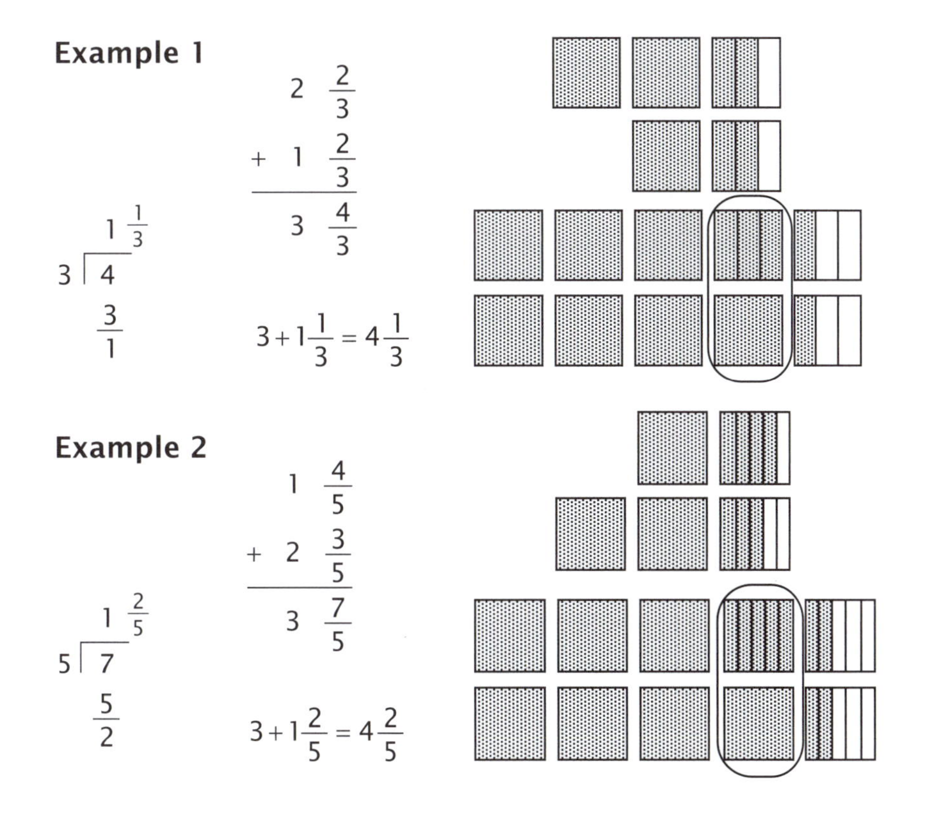

Example 3

$$\begin{array}{r} 2\ \frac{3}{4} \\ +\ \ \frac{3}{4} \\ \hline 2\ \frac{6}{4} \end{array}$$

$$4\overline{)6} = 1\frac{2}{4}$$
$$\begin{array}{r} 4 \\ \hline 2 \end{array}$$

$$2+1\frac{2}{4}=3\frac{2}{4}=3\frac{1}{2}$$

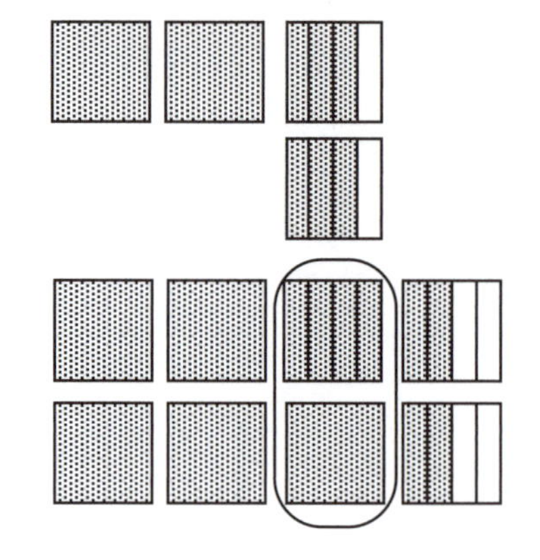

WORD PROBLEMS

Here are a few more multistep word problems to try. You may want to read and discuss these with your student as you work out the solutions together. Again, the purpose is to stretch rather than frustrate. If you do not think the student is ready, you may want to come back to these later.

The answers are at the end of the solutions section at the back of this book.

1. A rectangular field is 63 yards long and 21 yards wide. A fence is needed for the perimeter of the field. Fencing is also needed to divide the field into three square sections. How many feet of fencing are needed? (It is a good idea to make a drawing for this one.)

2. One-half of the people at the game wore the team colors. Two-thirds of those people wore team hats as well. One-fourth of those with team colors and team hats had banners to wave. Twenty-five people had team colors and banners, but not hats. One hundred people had only banners. If there were 1,824 people at the game, how many had banners?

3. A cube-shaped pool is half full of water. If the water is 36 inches deep, how much would the water in the pool weigh if the pool were filled to the brim? (One cubic foot of water weighs 56 pounds.)

LESSON 19

Subtraction of Mixed Numbers

with Regrouping

The inverse of adding mixed numbers is subtracting mixed numbers. Now, you can't subtract 2/3 from 1/3, so you need to borrow one from four, leaving three. Then change one to 3/3 so you can combine it with 1/3 to make 4/3.

Example 1

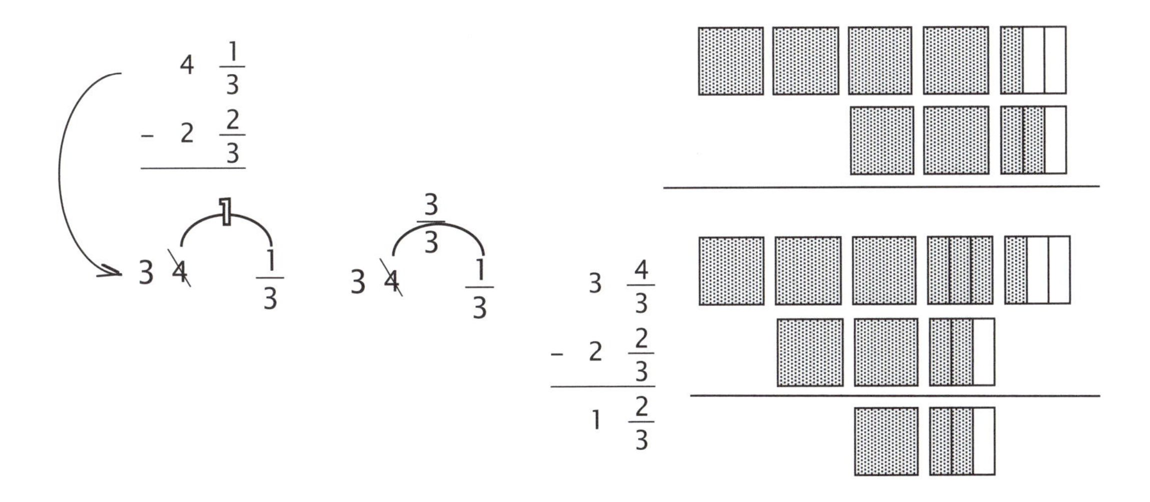

Example

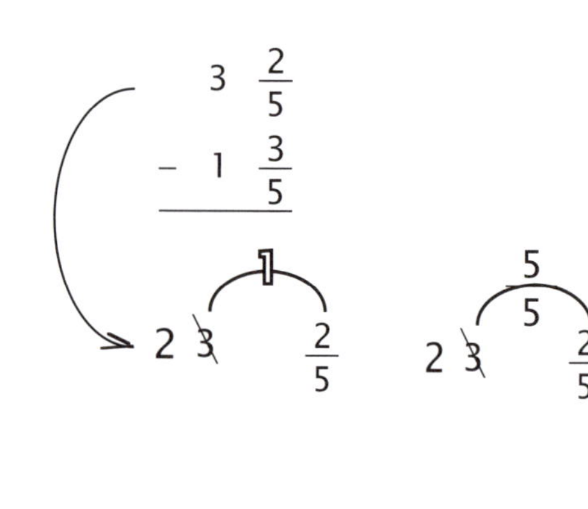

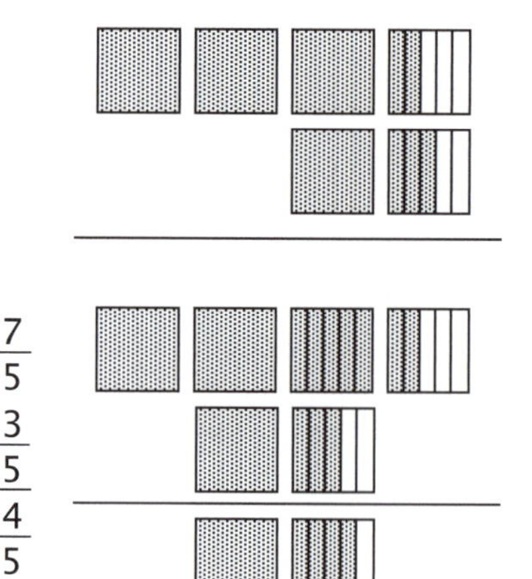

$$2\,\frac{7}{5} - 1\,\frac{3}{5} = 1\,\frac{4}{5}$$

LESSON 20

Subtraction of Mixed Numbers

with the Same Difference Theorem

The other method shown on the DVD, the same difference theorem, I recommend teaching after the student has learned the material taught in lesson 19. Look this lesson over, then redo the examples with this method and compare your answers.

Example 1 (from example 1, lesson 19)

$$\begin{array}{rcr} 4\frac{1}{3} + \frac{1}{3} & = & 4\frac{2}{3} \\ -\ 2\frac{2}{3} + \frac{1}{3} & = & -\ 3 \\ \hline & & 1\frac{2}{3} \end{array}$$

Here we ask, "What do I have to add to 2/3 to make this a **whole number**?" The answer is 1/3.

Then we have to add 1/3 to 4 1/3, as well, to make 4 2/3.

The problem is now 4 2/3 minus 3, which is much easier to subtract.

Example 2 (from example 2, lesson 19)

$$\begin{array}{rcr} 3\frac{2}{5} + \frac{2}{5} = & & 3\frac{4}{5} \\ -\ 1\frac{3}{5} + \frac{2}{5} = & & -\ 2 \\ \hline & & 1\frac{4}{5} \end{array}$$

Here we asked, "What do I have to add to 3/5 to make this a **whole number**?" The answer is 2/5.

Then we have to add 2/5 to 3 2/5, as well, to make 3 4/5.

The problem is now 3 4/5 minus 2, which is much easier to subtract.

LESSON 21

Adding Mixed Numbers

with Regrouping and Unequal Denominators; Mental Math

The only thing that is new in this lesson is that the denominators in the problems are not the same. These are as tough as mixed number problems get. In the examples, we take the fractions aside, get the same denominator, rewrite the problem, then perform the appropriate operation as in previous lessons.

Example 1

rule of four

$$1\frac{2}{3} + 5\frac{1}{4}$$

$$\frac{8}{12} + \frac{3}{12} = \frac{11}{12}$$

$$\begin{array}{r} 1\frac{8}{12} \\ +\ 5\frac{3}{12} \\ \hline 6\frac{11}{12} \end{array}$$

Example 2

rule of four

$$4\frac{3}{5} + 1\frac{1}{2}$$

$$\frac{6}{10} + \frac{5}{10} = \frac{11}{10}$$

$$\begin{array}{r} 4\frac{6}{10} \\ +\ 1\frac{5}{10} \\ \hline 5\frac{11}{10} \end{array} = 5\frac{10}{10} + \frac{1}{10}$$

$$= 5 + 1\frac{1}{10} = 6\frac{1}{10}$$

Example 3

rule of four

$$\begin{array}{r} 16\ \frac{4}{11} \\ +\ 23\ \frac{5}{7} \\ \hline \end{array} \qquad \frac{28_4}{77^{11}} + \frac{5^{55}}{7_{77}} = \frac{83}{77}$$

$$\begin{array}{r} 16\ \frac{28}{77} \\ +\ 23\ \frac{55}{77} \\ \hline 39\ \frac{83}{77} \end{array} = 39\frac{77}{77} + \frac{6}{77}$$

$$= 39 + 1\frac{6}{77} = 40\frac{6}{77}$$

MENTAL MATH

Here are some more mental math problems for you to read aloud to your student.

1. Thirty-six divided by six, plus three, times eight, equals ? (72)
2. Seventeen minus nine, times three, plus two, equals ? (26)
3. Three plus four, times six, minus three, equals ? (39)
4. Four times five, divided by 10 times seven, equals ? (14)
5. Sixty-six minus two, divided by eight, plus nine, equals ? (17)
6. Twenty-seven divided by nine, plus five, times seven, equals ? (56)
7. Thirty-one plus one, divided by eight, times six, equals ? (24)
8. Thirteen minus eight, times six, plus seven, equals ? (37)
9. Three times seven, plus three, divided by eight, equals ? (3)
10. Four times seven, plus eight, divided by four, equals ? (9)

Beginning with this lesson, there are also mental math problems on most of the systematic review pages in the *Epsilon* student text.

LESSON 22

Subtracting Mixed Numbers

with Regrouping and Unequal Denominators

In the examples, we take the fractions aside, get the same denominator, rewrite the problem, then perform the appropriate operation as in previous lessons. If we need to, we borrow (regroup) before subtracting.

Example 1

rule of four

$$\begin{array}{r} 3\ \frac{4}{5} \\ -\ \ \frac{1}{9} \\ \hline \end{array} \qquad \frac{36}{45} - \frac{5}{45} = \frac{31}{45} \qquad \begin{array}{r} 3\ \frac{36}{45} \\ -\ \ \frac{5}{45} \\ \hline 3\ \frac{31}{45} \end{array}$$

Example 2

rule of four

$$\begin{array}{r} 3\ \frac{2}{3} \\ -\ 1\ \frac{7}{8} \\ \hline \end{array} \qquad \frac{16}{24} - \frac{21}{24} \qquad \begin{array}{r} 3\ \frac{16}{24} \\ -\ 1\ \frac{21}{24} \\ \hline \end{array} \rightarrow \begin{array}{r} \overset{1 \text{ or } \frac{24}{24}}{2\not{3}\ \frac{40}{24}} \\ -\ 1\ \frac{21}{24} \\ \hline 1\ \frac{19}{24} \end{array}$$

Example 3 is done two ways. The first way is the traditional method, and the second way employs the same difference theorem. Use whichever method you feel most comfortable doing.

Example 3

rule of four

$$\begin{array}{r} 16\ \frac{1}{10} \\ -\ 3\ \frac{5}{7} \\ \hline \end{array} \qquad \frac{7_{1}}{70^{10}} - \frac{5^{50}}{7_{70}}$$

$$\begin{array}{r} 16\ \frac{7}{70} \\ -\ 3\ \frac{50}{70} \\ \hline \end{array} \longrightarrow \begin{array}{r} \overset{1 \text{ or } \frac{70}{70}}{15\not{16}}\ \frac{77}{70} \\ -\ 3\ \frac{50}{70} \\ \hline 12\ \frac{27}{70} \end{array}$$

The question is, "What do I have to add to 50/70 to make it a whole number?" The answer is 20/70. Adding this to both fractions produces 16 27/70 - 4, which equals 12 27/70.

$$\begin{array}{r} 16\ \frac{7}{70} \\ -\ 3\ \frac{50}{70} \\ \hline \end{array} \quad \begin{array}{r} +\frac{20}{70} \\ +\frac{20}{70} \end{array} \quad \begin{array}{r} 16\frac{27}{70} \\ -4 \\ \hline 12\frac{27}{70} \end{array}$$

LESSON 23

Dividing Fractions and Mixed Numbers

with the Reciprocal

Multiplying by the reciprocal of a number is the same as dividing by the number or fraction. A *reciprocal* is formed when the numerator and denominator switch places. The reciprocal of 2/3 is 3/2. The reciprocal of 4/11 is 11/4 (example 1). This skill is valuable in algebra because when you multiply a fraction by its reciprocal, the answer is one.

This method of dividing is often referred to as "invert and multiply." I prefer telling students that dividing by a fraction is equivalent to multiplying by the reciprocal of that fraction. Here is why division can be performed by multiplication of the reciprocal.

Example 1

$$\frac{2}{3} \times \frac{3}{2} = \frac{6}{6} = 1$$

$$\frac{4}{11} \times \frac{11}{4} = \frac{44}{44} = 1$$

Example 2

division $8 \div 2 = 4$

multiplication by the reciprocal $\frac{8}{1} \div \frac{2}{1} = \frac{8}{1} \times \frac{1}{2} = \frac{8}{2} = 4$

In the first method, eight is divided by two to get four. In the second method, eight is multiplied by the reciprocal of two, which is one-half. The result is four. Both methods produce the same answers.

Up to this point we have been dividing by making the denominators the same and dividing the numerators. This method will always work; we are just adding a second option. The following examples show how division can be done using both options.

Example 3

Solve $\frac{4}{5} \div \frac{1}{3}$.

Option 1 $\quad \frac{4}{5} \div \frac{1}{3} = \frac{12}{15} \div \frac{5}{15} = \frac{12}{5} = 2\frac{2}{5}$

Option 2 $\quad \frac{4}{5} \div \frac{1}{3} = \frac{4}{5} \times \frac{3}{1} = \frac{12}{5} = 2\frac{2}{5}$

There are advantages to both methods. Option 1 makes it easier to estimate the answer, but option 2 is often quicker.

When dividing mixed numbers, convert to improper fractions, then follow the same procedure.

Example 4

Solve $1\frac{1}{3} \div \frac{2}{5}$.

Option 1 $\quad 1\frac{1}{3} \div \frac{2}{5} = \frac{4}{3} \div \frac{2}{5} = \frac{20}{15} \div \frac{6}{15} = \frac{20}{6} = 3\frac{2}{6} = 3\frac{1}{3}$

Option 2 $\quad 1\frac{1}{3} \div \frac{2}{5} = \frac{4}{3} \div \frac{2}{5} = \frac{4}{3} \times \frac{5}{2} = \frac{20}{6} = 3\frac{2}{6} = 3\frac{1}{3}$

LESSON 24

Solving for an Unknown

with the Multiplicative Inverse; Word Problems

What do we multiply by 3/4 to make one? The answer is 4/3, because this is the reciprocal of 3/4, and 3/4 x 4/3 = 1. What do we multiply by five to make one? The answer is 1/5, because 1/5 is the reciprocal of 5/1, and 1/5 x 5/1 = 1. The *multiplicative inverse* of 5/1 is 1/5, since it is the inverse used in multiplying to produce one. The multiplicative inverse of 3/7 is 7/3. We use the multiplicative inverse to make one.

This concept of making one is necessary when solving an equation with a *coefficient* (a number in front of the variable or unknown), as the 3 in 3X or the 5 in 5Y. The primary objective when solving for an unknown with a coefficient is to get the variable with a coefficient of one. This will make it appear to have no coefficient, as $1 \cdot X$ is X. Study the following examples.

Example 1

$$3X = 12$$

$\frac{1}{3}$ times 3 equals $\frac{3}{3}$, which is 1.

$$\frac{1}{3} \cdot \frac{3}{1} X = \frac{12}{1} \cdot \frac{1}{3}$$

$$X = \frac{12}{3} = 4$$

$$\frac{1}{3} \times 3 = \frac{1}{3} \times \frac{3}{1} = \frac{3}{3} = 1$$

We multiply both sides by 1/3, the reciprocal of 3, or the multiplicative inverse, to get X all alone.

Example 2

$$5A = 35 \qquad \frac{1}{5} \text{ times 5 equals } \frac{5}{5}\text{, which is 1.}$$

$$\frac{1}{5} \cdot \frac{5}{1}A = \frac{35}{1} \cdot \frac{1}{5} \qquad \frac{1}{5} \times 5 = \frac{1}{5} \times \frac{5}{1} = \frac{5}{5} = 1$$

$$A = \frac{35}{5} = 7$$

We multiply both sides by 1/5, the reciprocal of 5, or the multiplicative inverse, to get A all alone.

Always check your work by replacing the letter (the unknown) with the number (the solution). Remember our *primary objective:* to find the value of X (or whatever letter is being used to represent the unknown) that will satisfy the equation. Here are checks for the previous examples. The solutions are placed in parentheses.

Example 1

check:

$$3X = 12$$
$$3(4) = 12$$
$$12 = 12$$

Example 2

check:

$$5A = 35$$
$$5(7) = 35$$
$$35 = 35$$

WORD PROBLEMS

Here are a few more multistep word problems to try. You may want to read and discuss these with your student as you work out the solutions together. Again, the purpose is to stretch, not to frustrate. If you do not think the student is ready, you may want to come back to these later.

The answers are at the end of the solutions section at the back of this book.

1. Mom mixed 2 1/2 pounds of apples, 1 1/8 pounds of grapes, and 1 1/4 pounds of pears for a salad. After setting aside 1 1/2 pounds of salad for today, she divided the rest of the salad equally into three containers. What is the weight of the salad in one container?

2. Peter wants to put five fish in his aquarium. Three of the fish need one-fourth of a cubic foot of water apiece, and two of them need one-third of a cubic foot of water each. The dimensions of Peter's aquarium are 1' x 1' x 2'. Does he have room to add another fish that needs two-thirds of a cubic foot of water? Assume that the aquarium is filled to the top.

3. Debbie had 5 1/2 yards of ribbon. She cut it into pieces that were each 1 1/2 yards long. How many inches long is the leftover piece? Note that the fraction that indicates the leftover piece is a fraction of a piece, not a fraction of the original amount.

LESSON 25

Multiplying Three Fractions

and Mixed Numbers

When multiplying fractions, you can find the answer two different ways. You can multiply the numerator and the denominator then reduce the answer if necessary, as in example 1. A shortcut method is to reduce the fraction in the middle of the problem instead of at the end. This is often referred to as canceling. In example 2 we'll do the problem in example 1 again, using this shortcut. Remember also that multiplication is commutative and the order of the factors may be changed without affecting the product.

Example 1

$$\frac{2}{3} \times \frac{5}{8} = \frac{10 \div 2}{24 \div 2} = \frac{5}{12}$$

In example 2 we can divide the two in the numerator by two to get one, and divide the eight in the denominator by two to get four. Then we multiply one by five to get five, and multiply three by four to get 12.

Example 2

$$\frac{2}{3} \times \frac{5}{8} = \frac{\overset{1}{\cancel{2}} \times 5}{3 \times \cancel{8}_{4}} = \frac{1 \times 5}{3 \times 4} = \frac{5}{12}$$

Both methods produce the same answer. Example 1 multiplies, then divides by a common factor of two to reduce. Example 2 divides by two in the middle of the problem, then multiplies for the answer.

We can divide by a common factor in the middle of the equation (canceling) or at the end of the equation (reducing). Another way to think of it is to rearrange the common factors on top of one another, because a number divided by itself is one. Notice this in example 3 with three above a three, and five above a five. These both will equal one.

Example 3

$$\frac{2}{3} \times \frac{5}{8} \times \frac{3}{5} = \frac{2 \times 5 \times 3}{3 \times 8 \times 5} = \frac{2 \times 3 \times 5}{8 \times 3 \times 5} = \frac{\cancel{2}^{1} \times \cancel{3}^{1} \times \cancel{5}^{1}}{\cancel{8}_{4} \times \cancel{3}_{1} \times \cancel{5}_{1}} = \frac{1}{4}$$

Now that we recognize that this works, we can solve example 3 without rearranging the factors, simply by dividing the numerator and denominator by a common factor. Example 4 is example 3 worked out in this manner step by step. You can do all the canceling without rewriting each step.

Example 4

$$\frac{2}{3} \times \frac{5}{8} \times \frac{3}{5} = \frac{2}{{}_{1}\cancel{3}} \times \frac{5}{8} \times \frac{\cancel{3}^{1}}{5} = \frac{2}{1} \times \frac{\cancel{5}^{1}}{8} \times \frac{1}{\cancel{5}_{1}}$$

$$= \frac{\cancel{2}^{1}}{1} \times \frac{1}{{}_{4}\cancel{8}} \times \frac{1}{1} = \frac{1}{1} \times \frac{1}{4} \times \frac{1}{1} = \frac{1}{4}$$

Another option is to multiply all the factors in the numerator and denominator, then reduce the final answer, as in example 5 below.

Example 5

$$\frac{2}{3} \times \frac{5}{8} \times \frac{3}{5} = \frac{30 \div 30}{120 \div 30} = \frac{1}{4}$$

Example 6

$$1\frac{1}{3}\times 2\frac{1}{2}\times\frac{3}{5}=\frac{4}{{}_{1}\cancel{3}}\times\frac{5}{2}\times\frac{\cancel{3}^{1}}{5}=\frac{4}{1}\times\frac{\cancel{5}^{1}}{2}\times\frac{1}{{}_{1}\cancel{5}}=$$

$$=\frac{\cancel{4}^{2}}{1}\times\frac{1}{{}_{1}\cancel{2}}\times\frac{1}{1}=\frac{2}{1}=2$$

or

$$1\frac{1}{3}\times 2\frac{1}{2}\times\frac{3}{5}=\frac{\cancel{4}^{2}}{{}_{1}\cancel{3}}\times\frac{\cancel{5}^{1}}{{}_{1}\cancel{2}}\times\frac{\cancel{3}^{1}}{{}_{1}\cancel{5}}=\frac{2}{1}=2$$

or

$$1\frac{1}{3}\times 2\frac{1}{2}\times\frac{3}{5}=\frac{4}{3}\times\frac{5}{2}\times\frac{3}{5}=\frac{60\ \div\ 30}{30\ \div\ 30}=\frac{2}{1}=2$$

Example 7

$$1\frac{1}{5}\times 3\frac{1}{3}\times 1\frac{3}{4}=\frac{6}{5}\times\frac{\cancel{10}^{5}}{3}\times\frac{7}{{}_{2}\cancel{4}}=\frac{6}{{}_{1}\cancel{5}}\times\frac{\cancel{5}^{1}}{3}\times\frac{7}{2}$$

$$=\frac{\cancel{6}^{2}}{1}\times\frac{1}{{}_{1}\cancel{3}}\times\frac{7}{2}=\frac{\cancel{2}^{1}}{1}\times\frac{1}{1}\times\frac{7}{{}_{1}\cancel{2}}=7$$

or

$$1\frac{1}{5}\times 3\frac{1}{3}\times 1\frac{3}{4}=\frac{{}^{1}\cancel{2}\cancel{6}}{{}_{1}\cancel{5}}\times\frac{\cancel{5}^{1}\cancel{10}}{{}_{1}\cancel{3}}\times\frac{7}{{}_{1}\cancel{2}\cancel{4}}=7$$

or

$$1\frac{1}{5}\times 3\frac{1}{3}\times 1\frac{3}{4}=\frac{6}{5}\times\frac{10}{3}\times\frac{7}{4}=\frac{420\ \div\ 60}{60\ \div\ 60}=\frac{7}{1}=7$$

Notice that you can keep canceling as long as you have a number in the numerator that is a factor of a number in the denominator, or a number in the denominator that is a factor of a number in the numerator.

In the third line of example 7, the six is divided by three, which leaves two in the numerator. Then the four in the denominator and 10 in the numerator are both divided by two, which leaves two in the denominator and five in the numerator. We still have a two in the numerator and a two in the denominator. We can divide both again by two. Even though it is called canceling, the numbers, when divided by a common factor, don't disappear; rather, they are absorbed into the final answer.

LESSON 26

Solving for an Unknown

with Multiplicative and Additive Inverses

In lesson 24, we learned about the multiplicative inverse, which is the inverse used in multiplying to produce one. The *additive inverse* is used in adding and subtracting to make zero. The additive inverse of adding two is subtracting two. The additive inverse of subtracting 13 is adding 13. The opposite, or inverse, of adding is subtracting.

When we have a number added to an equation, we can subtract the same number from both sides so the variable will be by itself on one side of the equation.

Example 1 below begins with $2X + 3 = 11$. What we are solving for is the value of X that makes both sides equal, or the same. We start by subtracting three from both sides of the equation, which results in $2X = 8$. Then we multiply both sides by the reciprocal, which is 1/2. The equation then reads $X = 4$. To check if this is the correct answer, we put the 4 in place of the X in the original equation to see if it makes both sides equal.

Example 1

$$2X + 3 = 11$$

$$\begin{array}{r} 2X + 3 = 11 \\ \underline{-3} \;\; \underline{-3} \end{array}$$

Subtract 3 from both sides.

$$\frac{2}{4} \cdot 2X = 8 \cdot \frac{1}{2}$$

Multiply both sides by the reciprocal of 2, which is 1/2.

$$X = 4$$

Always check your work by replacing the letter (the unknown) with the number (the solution). Remember our *primary objective*: to find the value of X (or whatever letter is being used to represent the unknown) that will satisfy the equation. The solution is placed in parentheses.

Example 1

check:

$$2X + 3 = 11$$
$$2(4) + 3 = 11$$
$$8 + 3 = 11$$
$$11 = 11$$

When a number is subtracted from an equation, we add the same number to both sides of the equation.

Example 2

$$5X - 13 = 2$$

$$\begin{array}{r} 5X - 13 = 2 \\ \underline{+13 \; +13} \end{array}$$

Add 13 to both sides.

$$\frac{1}{5} \cdot 5X = 15 \cdot \frac{1}{5}$$

Multiply both sides by the reciprocal of 5, which is $\frac{1}{5}$.

$$X = 3$$

Example 2

check:

$$5X - 13 = 2$$
$$5(3) - 13 = 2$$
$$15 - 13 = 2$$
$$2 = 2$$

LESSON 27

Area and Circumference of Circle

with Pi = 22/7

The formula for the area of a circle is πr^2. Pi, or π, is the symbol for a value that is a little more than three. The r represents the radius of a circle. The *radius* is the distance from the center of the circle to the edge of the circle. To help you understand and remember the value of π, see figure 1 below. When finding area, remember the word "squarea," which is a conjoining of the words *square* and *area*. Area is always computed in square units. The value of π is normally represented by one of two values, 22/7 or 3.14. Both of these are approximations for a number that extends as a decimal indefinitely. (Pi = 3.1415927. . .) In some problems it is advantageous to use the decimal, while in others the fraction is more convenient. But since the focus of this book is fractions, we'll use 22/7 in all of our problems.

Figure 1

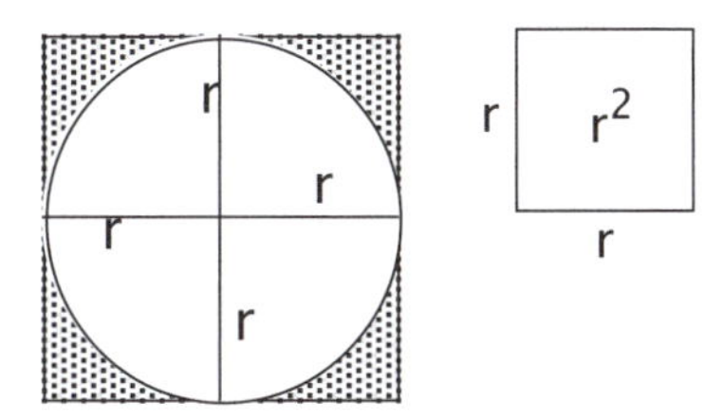

A little more than three squares with a side of length r and an area of r^2 is the area of a circle. When a number has a small 2 above it, as in r^2, it indicates that you use the number as a factor two times.

3^2 is the same as 3 x 3 or 9.

$5^2 = 5 \times 5 = 25$.

Example 1

Find the area of the circle.

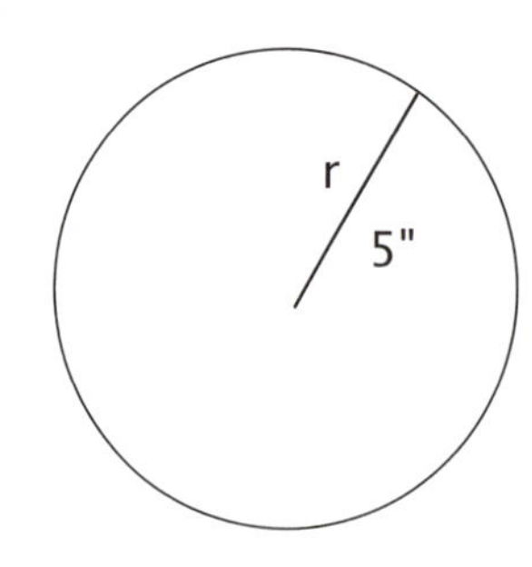

Area = πr^2

Area = $(\frac{22}{7})(5^2)$

Area = $\frac{22}{7} \times \frac{25}{1} = \frac{550}{7} = 78\frac{4}{7}$ square inches

Example 2

Find the area of the circle.

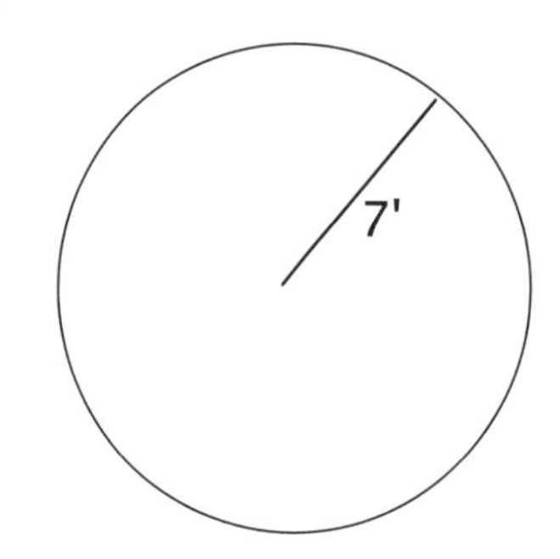

Area = πr^2

Area = $(\frac{22}{7})(7^2)$

Area = $\frac{22}{7} \times \frac{49}{1} = \frac{22}{\cancel{7}_1} \times \frac{\cancel{49}^7}{1} = \frac{154}{1} = 154$ square feet

CIRCUMFERENCE OF A CIRCLE

Rectangles have area and perimeter. Circles have area and *circumference*. The circumference is the distance around the outside of a circle. A practical example of this is the length of a belt, which goes around your waist. Circumference is always computed in linear units such as plain inches or plain feet. The formula for the circumference of a circle is $2\pi r$. The r represents the radius of a circle. The value of π is normally represented by by one of two values, 22/7 or 3.14. As with area, we'll use 22/7 for π in all of our problems.

Example 1

Find the circumference of the circle.

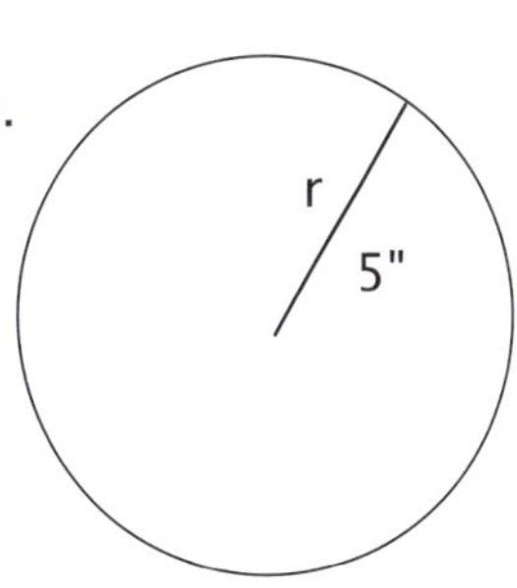

Circumference = $2\pi r$

Circumference = $2(\frac{22}{7})(5")$

Circumference = $\frac{2}{1} \times \frac{22}{7} \times \frac{5}{1} = \frac{220}{7} = 31\frac{3}{7}$ inches

Example 2

Find the circumference of the circle.

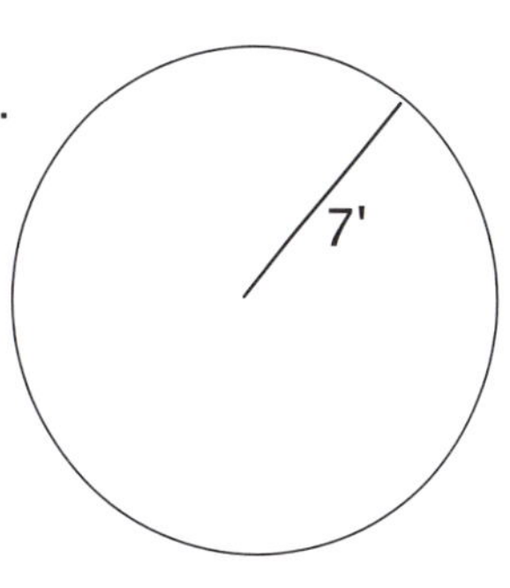

Circumference = $2\pi r$

Circumference = $2(\frac{22}{7})(7")$

Circumference = $\frac{2}{1} \times \frac{22}{7} \times \frac{7}{1} = \frac{308}{7} = 44$ feet

or

$= \frac{2}{1} \times \frac{22}{\cancel{7}_1} \times \frac{\cancel{7}^1}{1} = \frac{44}{1} = 44$ feet

LESSON 28

Solving for an Unknown

with Fractional Coefficients

This lesson is very similar to lesson 26, except that here the coefficients are fractions instead of whole numbers. After you solve for the unknown, check your answer as shown in the following the examples.

Example 1

$$\frac{2}{3}X = 12$$

$$\frac{3}{2} \cdot \frac{2}{3}X = 12 \cdot \frac{3}{2}$$

$$X = 18$$

$$\frac{3}{2} \times \frac{2}{3} = \frac{6}{6} = 1$$

$$\frac{3}{2} \times \frac{12}{1} = \frac{36}{2} = 18$$

We multiply both sides by 3/2,
the reciprocal of 2/3, to get X all alone.

Example 1 (check)

$$\frac{2}{3}X = 12$$

$$\frac{2}{3}(18) = 12 \longrightarrow \frac{2}{3} \times \frac{18}{1} = \frac{36}{3} = 12$$

$$12 = 12$$

Always check your work by replacing the letter (the unknown) with the number (the solution). Remember our primary objective: to find the value of X (or whatever letter is being used to represent the unknown) that will satisfy the equation.

Example 2

$$\frac{1}{2}X + 3 = 11$$
$$\quad -3 \quad -3$$
$$\frac{2}{1} \cdot \frac{1}{2}X = 8 \cdot \frac{2}{1}$$
$$X = 16$$

$$\frac{2}{1} \times \frac{1}{2} = \frac{2}{2} = 1$$
$$\frac{2}{1} \times \frac{8}{1} = \frac{16}{1} = 16$$

We multiply both sides by 2/1, the reciprocal of 1/2, to get X all alone.

Example 2 (check)

$$\frac{1}{2}X + 3 = 11$$
$$\frac{1}{2}(16) + 3 = 11$$
$$8 + 3 = 11$$
$$11 = 11$$

$$\longrightarrow \frac{1}{\cancel{2}_1} \times \frac{\cancel{16}^{8}}{1} = \frac{8}{1} = 8$$

LESSON 29

Fractions to Decimals to Percents

A *decimal* is a fraction written on one line (without a numerator or denominator) in the decimal system. In the decimal system there are values for each place. This is called *place value* (see figure 1).

Figure 1

___	___	___	___	.	___	___	___
1,000	100	10	1		$\frac{1}{10}$	$\frac{1}{100}$	$\frac{1}{1,000}$

All numbers in the decimal system have a digit (0 through 9) and a place value. The number 142 may be written in *expanded notation* as 1x100 + 4x10 + 2x1. The digits are 1, 4, and 2, and the values are hundreds, tens, and units (ones). In figure 2 below, 142 is written in the decimal system.

Figure 2

___	1	4	2	.	___	___	___
1,000	100	10	1		$\frac{1}{10}$	$\frac{1}{100}$	$\frac{1}{1,000}$

The number .5 is called a decimal number. Some call it a decimal fraction. It is five-tenths and is written in expanded notation as 5 x 1/10. In figure 3, we see .5 written in the decimal system.

Figure 3

___	___	___	___	.	5	___	___
1,000	100	10	1		$\frac{1}{10}$	$\frac{1}{100}$	$\frac{1}{1,000}$

A fraction can also be written in expanded notation. Two-fifths (2/5) is 2 x 1/5. We can't write 2/5 as a decimal because there is no fifths place in the decimal system. We must change the denominator to 10, 100, or 1,000, or some other multiple of ten. To show this, build 2/5, then place the 1/10 overlay on top of it as shown in example 1.

Example 1

Change 2/5 to a decimal number.

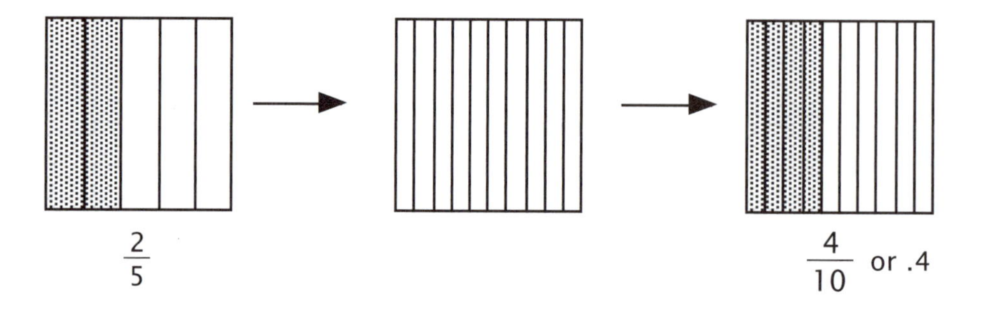

We have just transformed 2/5 to 4/10, which may be expressed as .4, or four-tenths in the decimal system.

Example 2

Change 1/2 to a decimal number.

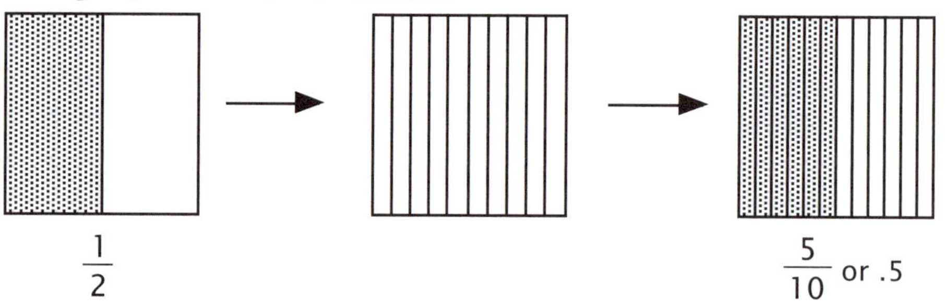

The key is to change the denominator to a power of 10 so that it can be written in the decimal system.

Example 3

Change $\frac{3}{4}$ to a decimal number.

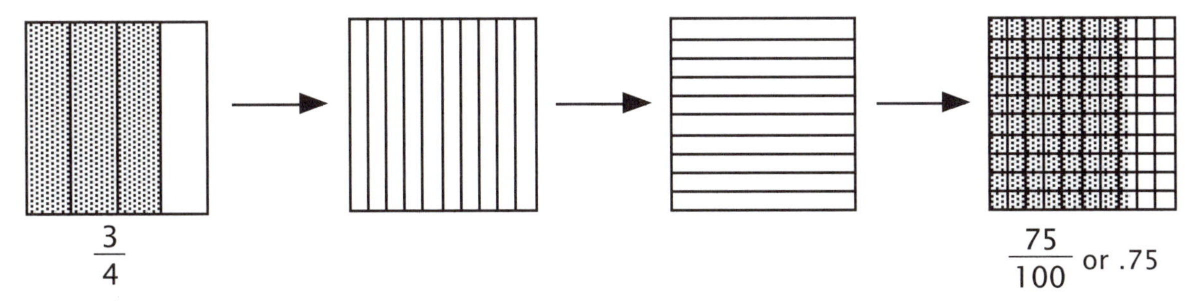

To illustrate this problem clearly, make three-fourths with the fraction kit, then take off the clear fourths overlay. Place the two tenths overlays on top of three-fourths with one overlay turned to make a 100 grid. Three-fourths does not transform into tenths evenly, so we have transformed three-fourths into hundredths. You can see seven rows of 10 for 70, then a half row of 10, or five, to make 70 + 5 or 75.

The following fractions can be used to illustrate changing a fraction to a decimal. They work out exactly. Other fractions, such as the thirds and the sixths, don't come out evenly, but they still reveal a fine approximation.

$$\frac{1}{5}, \frac{2}{5}, \frac{3}{5}, \frac{4}{5}, \frac{5}{5}, \frac{1}{2}, \frac{1}{4}, \frac{3}{4}$$

To change a fraction or a decimal to a percent, recognize that *percent* means "per hundred." First transform a fraction or decimal to hundredths (1/100) as in example 3. Then watch the progression in example 4.

Example 4

Change 3/4 and .75 to a percent.

$$\frac{3}{4} = .75 = \frac{75}{100} \longrightarrow \frac{75\,/}{00} \longrightarrow \frac{75\%}{0} \longrightarrow \underline{75\%} \longrightarrow 75\%$$

Beginning with the 100 in the denominator of 75/100, take the one and the two zeros, and step by step, use them to form the percent symbol. This method clearly illustrates how percents are a different way of symbolizing hundredths. Once a fraction or decimal is changed to hundredths, transitioning to percent is quick and easy.

LESSON 30

More Solving for an Unknown

In lesson 28 we learned about solving for an unknown with multiplicative and additive inverses when the coefficient of the unknown was a fraction. The only thing new in this lesson is that the quantity to be added to both sides is a fraction, not a whole number. When you multiply by the reciprocal of 2/3, you are using a multiplicative inverse. When you add or subtract 2/3, you are using an additive inverse. See example 1.

Example 1

$$\frac{3}{5}G - \frac{1}{2} = 5\frac{1}{2}$$

$$+\frac{1}{2} \quad +\frac{1}{2}$$

Add the opposite of (-1/2), which is (+1/2), to both sides.

$$\frac{5}{3} \cdot \frac{3}{5}G = 6 \cdot \frac{5}{3}$$

Multiply both sides by 5/3, the reciprocal of 3/5.

$$G = 10$$

$$\frac{5}{3} \times \frac{6}{1} = 10$$

Example 1

check

$$\frac{3}{5}G - \frac{1}{2} = 5\frac{1}{2}$$

$$\frac{3}{5}(10) - \frac{1}{2} = 5\frac{1}{2} \longrightarrow \frac{3}{5} \times \frac{10}{1} = 6$$

$$6 - \frac{1}{2} = 5\frac{1}{2}$$

$$5\frac{1}{2} = 5\frac{1}{2}$$

Example 2

$$\frac{1}{4}M + \frac{5}{8} = \frac{3}{4}$$
$$-\frac{5}{8} \quad -\frac{5}{8}$$
$$\frac{4}{1}\cdot\frac{1}{4}M = \frac{1}{8}\cdot\frac{4}{1}$$
$$M = \frac{1}{2}$$

Subtract 5/8 from both sides.

$$\frac{3}{4} - \frac{5}{8} = \frac{24}{32} - \frac{20}{32} = \frac{4}{32} = \frac{1}{8}$$

Multiply both sides by 4/1, the reciprocal of 1/4.

$$\frac{4}{1} \times \frac{1}{8} = \frac{4}{8} = \frac{1}{2}$$

Example 2

check

$$\frac{1}{4}\cdot\frac{1}{2} + \frac{5}{8} = \frac{3}{4}$$
$$\frac{1}{8} + \frac{5}{8} = \frac{3}{4}$$
$$\frac{3}{4} = \frac{3}{4}$$

APPENDIX A

Finding the Area of a Trapezoid

Finding the area of a trapezoid is taught in *Delta* and again in *Geometry*. This lesson is for those who started Math-U-See after the *Delta* level or those who are transferring from the classic *Intermediate*. It may also be used as a review for other students.

A *trapezoid* is a *quadrilateral* (a four-sided figure with at least one set of parallel sides). In the picture below, notice that the top and bottom sides are parallel, but the other two sides are not. The top and bottom are called the *bases*. Finding the area of a trapezoid is similar to finding the area of a rectangle. The area of a rectangle is found by multiplying the base by the height. The formula for the area of a trapezoid is the *average* base times the height. Consider figure 1 to see where this formula originates.

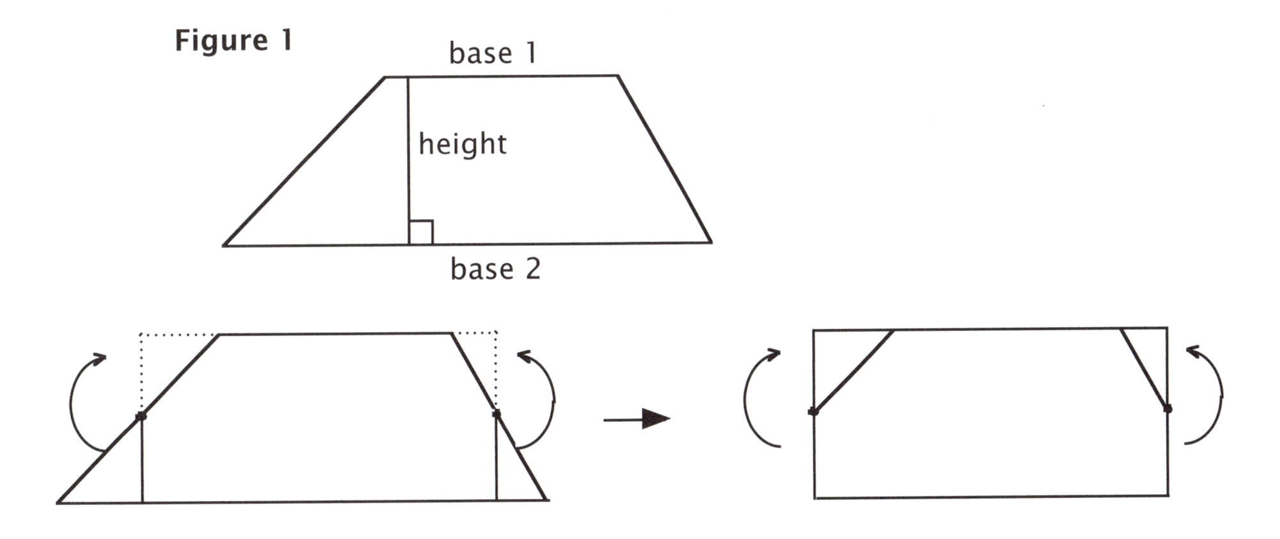

On the left and right sides in figure 1, we choose the point in the middle and make one small triangle on each side. Then we pivot on this midpoint and swing the lower triangles up on both sides to make a rectangle out of the trapezoid. The resulting base is the average of the top and bottom bases and is found by connecting the two midpoints on the sides.

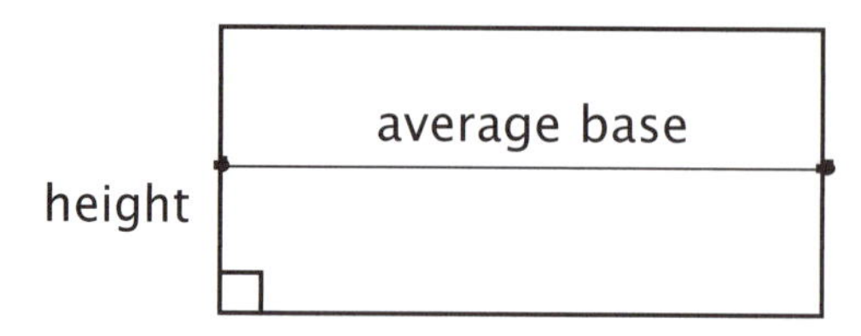

The traditional formula for finding the area of a trapezoid is $\frac{b_1 + b_2}{2} \times h$

The average base is found by adding the top and bottom bases and dividing by two. Then this result is multiplied by the height to find the area.

Example 1

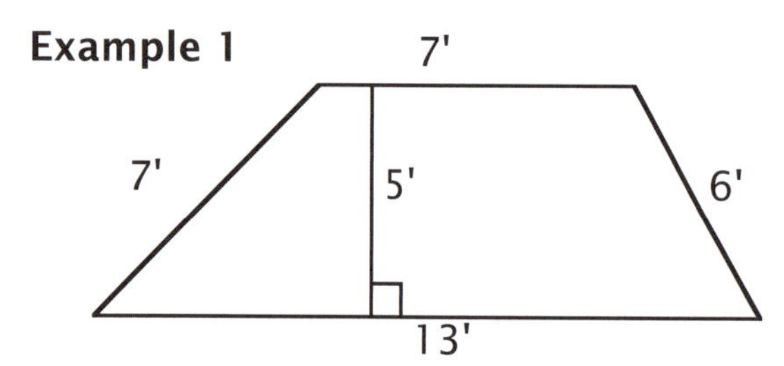

Find the area of the trapezoid.
The average base is (7 + 13) ÷ 2 = 10'.
The height is 5'.
The area is 10 x 5 = 50 square feet or 50 ft^2.

Example 2

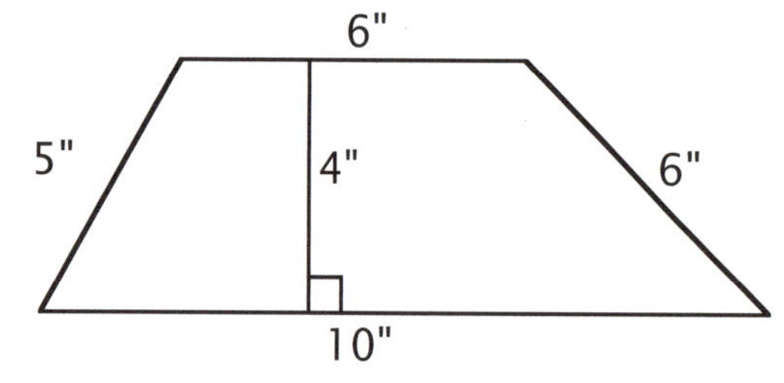

Find the area of the trapezoid.
The average base is (6 +10) ÷ 2 = 8".
The height is 4".
The area is 8" x 4" = 32 square inches or 32 in^2.

Student Solutions

Lesson Practice 1A

1. done
2. $6;2;1;\frac{1}{2}$ of 6 is 3
3. $10;5;4;\frac{4}{5}$ of 10 is 8
4. $9;3;2;\frac{2}{3}$ of 9 is 6
5. $8;4;3;\frac{3}{4}$ of 8 is 6

Lesson Practice 1B

1. $\frac{3}{5}$ of 15 is 9
2. $\frac{2}{3}$ of 18 is 12
3. $\frac{5}{8}$ of 8 is 5
4. $20 \div 5 = 4$
   $4 \times 3 = 12$
5. $6 \div 3 = 2$
   $2 \times 2 = 4$
6. $8 \div 2 = 4$
   $4 \times 1 = 4$
7. $4 \div 4 = 1$
   $1 \times 3 = 3$
8. $6 \div 3 = 2$
   $2 \times 1 = 2$
9. $8 \div 4 = 2$
   $2 \times 1 = 2$
10. $10 \div 5 = 2$
    $2 \times 4 = 8$
11. $6 \div 2 = 3$
    $3 \times 1 = 3$
12. $12 \div 3 = 4$
    $4 \times 2 = 8$
13. $10 \div 5 = 2$
    $2 \times 2 = 4$
14. $12 \div 4 = 3$
    $3 \times 1 = 3$ eggs
15. $10 \div 2 = 5$
    $5 \times 1 = 5$ boys

Lesson Practice 1C

1. $\frac{1}{3}$ of 12 is 4
2. $\frac{3}{4}$ of 20 is 15
3. $\frac{7}{9}$ of 18 is 14
4. $16 \div 4 = 4$
   4x2 = 8
5. $10 \div 5 = 2$
   2x1 = 2
6. $18 \div 6 = 3$
   3x5 = 15
7. $9 \div 3 = 3$
   3x1 = 3
8. $12 \div 4 = 3$
   3x3 = 9
9. $10 \div 5 = 2$
   2x3 = 6
10. $4 \div 2 = 2$
    2x1 = 2
11. $12 \div 3 = 4$
    4x1 = 4
12. $12 \div 2 = 6$
    6x1 = 6
13. $16 \div 8 = 2$
    2x7 = 14
14. $14 \div 7 = 2$
    2x3 = 6 days
15. $20 \div 5 = 4$
    4x4 = 16 tulips

Systematic Review 1D

1. $\frac{2}{5}$ of 15 is 6
2. $\frac{1}{3}$ of 9 is 3
3. $18 \div 9 = 2$
   $2 \times 5 = 10$
4. $10 \div 5 = 2$
   $2 \times 2 = 4$
5. $8 \div 4 = 2$
   $2 \times 3 = 6$
6. $30 \div 6 = 5$
   $5 \times 5 = 25$
7. $48 \div 8 = 6$
   $6 \times 7 = 42$
8. $24 \div 6 = 4$
   $4 \times 3 = 12$
9. $35 \div 7 = 5$
   $5 \times 4 = 20$
10. $10 \div 5 = 2$
    $2 \times 3 = 6$
11. done
12. $20+10+20+10 = 60$ ft
13. $15+7+15+7 = 44$ in
14. $60 \div 3 = 20$
    $20 \times 2 = 40$ minutes
15. $25 \div 5 = 5$
    $5 \times 3 = 15$ girls

Systematic Review 1E

1. $\frac{4}{5}$ of 20 is 16
2. $\frac{1}{2}$ of 6 is 3
3. $10 \div 5 = 2$
   $2 \times 5 = 10$
4. $20 \div 4 = 5$
   $5 \times 3 = 15$
5. $24 \div 6 = 4$
   $4 \times 5 = 20$
6. $56 \div 7 = 8$
   $8 \times 1 = 8$
7. $12 \div 3 = 4$
   $4 \times 2 = 8$
8. $10 \div 5 = 2$
   $2 \times 2 = 4$
9. $8 \div 4 = 2$
   $2 \times 1 = 2$
10. $32 \div 8 = 4$
    $4 \times 3 = 12$
11. $25+13+25+13 = 76$ in
12. $32+16+32+16 = 96$ ft
13. $78+39+78+39 = 234$ in
14. $\$16 \div 2 = \$8$
    $\$8 \times 1 = \$8$
15. $20+30+20+30 = 100$ ft
16. $56 \div 7 = 8$
    $8 \times 4 = 32$ plates
17. $6+11+6+11 = 34$ ft
18. $18 \div 9 = 2$
    $2 \times 7 = 14$ students

Systematic Review 1F

1. $\frac{3}{6}$ of 24 = 12
2. $\frac{2}{4}$ of 8 = 4
3. $21 \div 3 = 7$
   $7 \times 2 = 14$
4. $6 \div 3 = 2$
   $2 \times 1 = 2$
5. $12 \div 3 = 4$
   $4 \times 1 = 4$
6. $12 \div 4 = 3$
   $3 \times 3 = 9$
7. $25 \div 5 = 5$
   $5 \times 1 = 5$
8. $42 \div 6 = 7$
   $7 \times 5 = 35$
9. $54 \div 9 = 6$
   $6 \times 1 = 6$
10. $16 \div 8 = 2$
    $2 \times 5 = 10$

11. $9+6+9+6=30$ in
12. $18+10+18+10=56$ ft
13. $43+27+43+27=140$ in
14. $36 \div 9=4$
    $4 \text{x} 5=20$ candies
15. $36 \div 9=4$
    $4 \text{x} 4=16$
    $16+20=36$ candies
16. $\$12.00+\$20.00+\$8.50=\$40.50$
17. $\$40.50-\$14.50=\$26.00$
18. $12 \div 4=3$
    $3 \text{x} 3=9$ months

Lesson Practice 2A

1. done
2. $\frac{3}{5}$; three-fifths
3. $\frac{1}{5}$; one-fifth
4. $\frac{2}{6}$; two-sixths
5. $\frac{2}{3}$; two-thirds
6. $\frac{2}{2}$; two-halves
7. $\frac{3}{4}$; three-fourths
8. $\frac{4}{6}$; four-sixths
9. done
10. two-fourths; 4 sections, 2 shaded

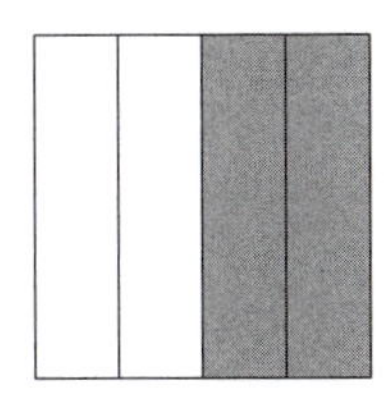

11. one-third; 3 sections, 1 shaded

12. two-halves; 2 sections, both shaded

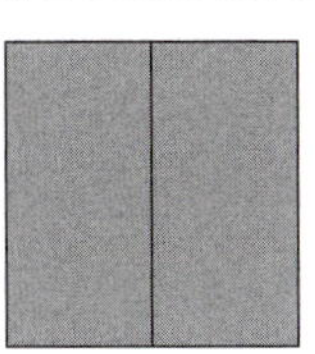

13. done
14. $\frac{2}{3}$; 3 sections, 2 shaded

15. $\frac{1}{4}$; 4 sections, 1 shaded

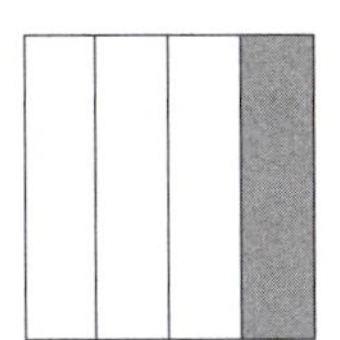

16. $\frac{3}{6}$; 6 sections, 3 shaded

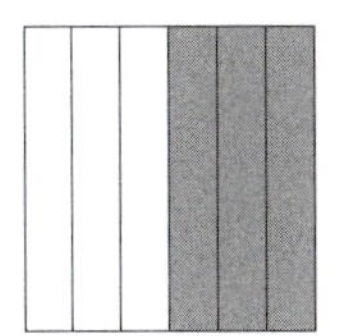

Lesson Practice 2B

1. $\frac{3}{3}$; three-thirds
2. $\frac{1}{6}$; one-sixth
3. $\frac{2}{5}$; two-fifths
4. $\frac{6}{6}$; six-sixths
5. $\frac{1}{3}$; one-third

6. $\frac{1}{2}$; one-half
7. $\frac{2}{4}$; two-fourths
8. $\frac{4}{5}$; four-fifths
9. three-thirds; 3 sections, all shaded

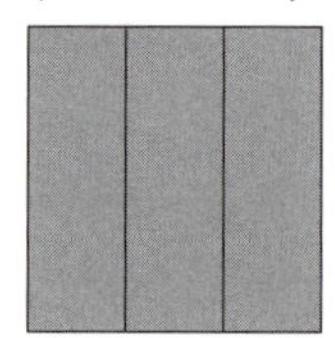

10. one-sixth; 6 sections, 1 shaded

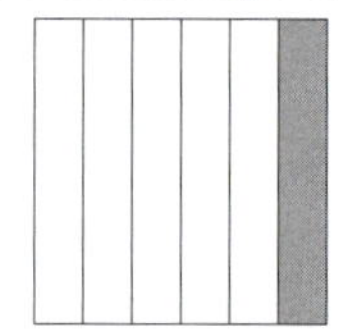

11. one-half; 2 sections, 1 shaded

12. four – sixths; 6 sections, 4 shaded

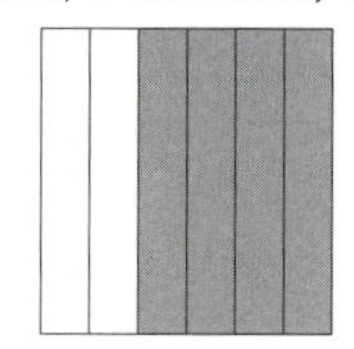

13. $\frac{3}{5}$; 5 sections, 3 shaded

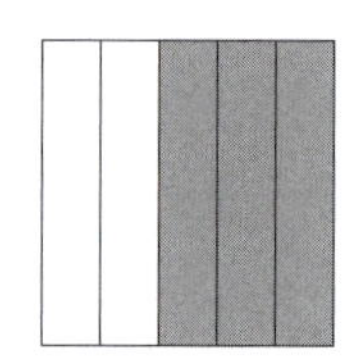

14. $\frac{4}{4}$; 4 sections, all shaded

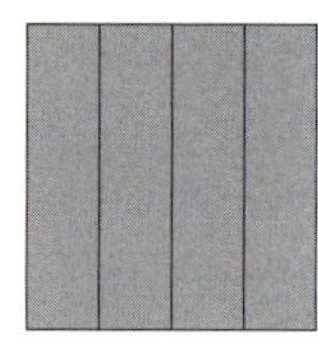

15. $\frac{4}{5}$; 5 sections, 4 shaded

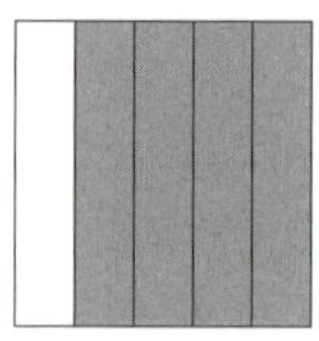

16. $\frac{5}{6}$; 6 sections, 5 shaded

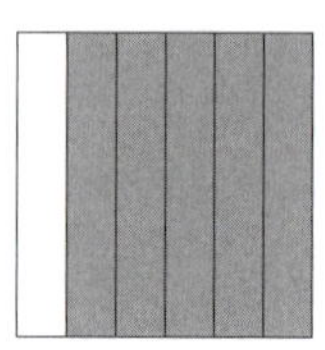

Lesson Practice 2C

1. $\frac{2}{3}$; two-thirds
2. $\frac{5}{6}$; five-sixths
3. $\frac{3}{5}$; three-fifths
4. $\frac{5}{6}$; five-sixths
5. $\frac{2}{2}$; two-halves
6. $\frac{1}{3}$; one-third
7. $\frac{3}{4}$; three-fourths
8. $\frac{1}{5}$; one-fifth
9. two-sixths; 6 sections, 2 shaded

10. one-fourth; 4 sections, 1 shaded

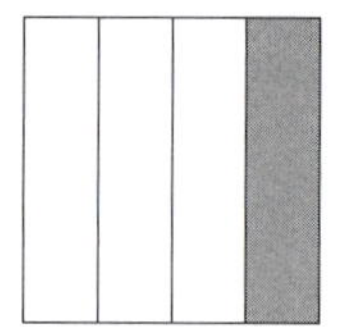

11. two-thirds; 3 sections, 2 shaded

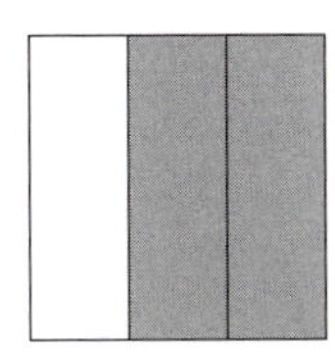

12. four – fifths; 5 sections, 4 shadec

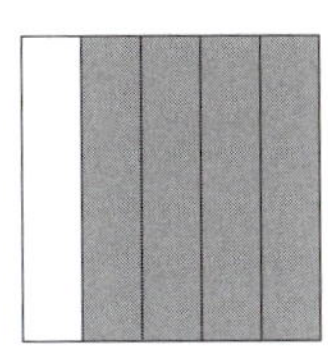

13. $\frac{1}{5}$; 5 sections, 1 shaded

14. $\frac{2}{4}$; 4 sections, 2 shaded

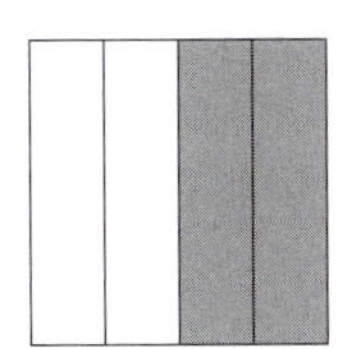

15. $\frac{1}{3}$; 3 sections, 1 shaded

16. $\frac{6}{6}$; 6 sections, all shaded

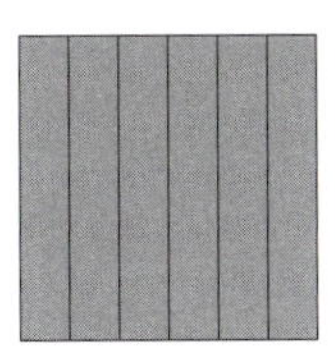

Systematic Review 2D

1. $\frac{1}{4}$; one-fourth
2. $\frac{5}{5}$; five-fifths
3. $\frac{4}{4}$; four-fourths
4. three – sixths; 6 sections, 3 shaded
5. one – fifth; 5 sections, 1 shaded
6. $\frac{5}{6}$; 6 sections, 5 shaded
7. $24 \div 3 = 8; 8x2 = 16$
8. $40 \div 4 = 10; 3x10 = 30$
9. $34 \div 2 = 17; 1x17 = 17$
10. done
11. $15 + 15 + 15 + 15 = 60$ ft
12. rectangle:
    $12 + 24 + 12 + 24 = 72$ in
    square:
    $10+10+10+10=40$ in
    combined:
    $72+40=112$ in
13. $18 \div 9 = 2$
    $2x5 = 10$ players
14. $12 \div 4 = 3$
    $3x3 = 9$ bushes

Systematic Review 2E

1. $\frac{3}{6}$; three-sixths
2. $\frac{5}{6}$; five-sixths
3. $\frac{2}{5}$; two-fifths
4. $\frac{1}{6}$; 6 sections, 1 shaded
5. $\frac{2}{3}$; 3 sections, 2 shaded
6. six – sixths; 6 sections, all shaded
7. $36 \div 6 = 6$
    $1x6 = 6$
8. $72 \div 9 = 8$
    $8x4 = 32$

9. $39 \div 3 = 13$
   $13 \times 2 = 26$
10. $4+17+4+17 = 42$ yd
11. $34+34+34+34 = 136$ ft
12. $18+18+18+18 = 72$ yd to start
    $72-10=62$ yd left
13. $12 \div 4 = 3$
    $3 \times 1 = 3$ months
14. $12 \div 3 = 4$
    $4 \times 2 = 8$ months
15. $35 \div 7 = 5$
    $5 \times 5 = 25$ camels

Systematic Review 2F

1. $\frac{2}{4}$; two-fourths
2. $\frac{4}{5}$; four-fifths
3. $\frac{2}{6}$; two-sixths
4. $\frac{3}{5}$; 5 sections, 3 shaded
5. $\frac{2}{6}$; six sections, 2 shaded
6. three-fourths; 4 sections, 3 shaded
7. $80 \div 2 = 40; 1 \times 40 = 40$
8. $55 \div 5 = 11; 11 \times 3 = 33$
9. $56 \div 8 = 7; 7 \times 3 = 21$
10. $46+21+46+21 = 134$ yd
11. $13+13+13+13 = 52$ ft
12. $12 \div 6 = 2$
    $2 \times 1 = 2$ eggs
13. $20 \div 5 = 4$
    $4 \times 3 = 12$ people
14. $20 \div 5 = 4$
    $4 \times 2 = 8$ people
15. $\$10.50 + \$7.25 = \$17.75$

Lesson Practice 3A

1. $\frac{5}{5}$; five-fifths
2. $\frac{1}{3}$; one-third
3. $\frac{1}{4}+\frac{2}{4}=\frac{3}{4}$ or three-fourths
4. $\frac{3}{6}-\frac{2}{6}=\frac{1}{6}$ or one-sixth
5. $\frac{1}{6}+\frac{3}{6}=\frac{4}{6}$
6. $\frac{1}{5}+\frac{2}{5}=\frac{3}{5}$
7. $\frac{4}{6}+\frac{1}{6}=\frac{5}{6}$
8. $\frac{4}{6}-\frac{1}{6}=\frac{3}{6}$
9. $\frac{3}{4}-\frac{2}{4}=\frac{1}{4}$
10. $\frac{3}{6}-\frac{2}{6}=\frac{1}{6}$
11. $\frac{1}{4}+\frac{2}{4}=\frac{3}{4}$ of the letters
12. $\frac{1}{5}+\frac{3}{5}=\frac{4}{5}$ of the book
13. $\frac{6}{6}-\frac{2}{6}=\frac{4}{6}$ of the job left
14. $\frac{4}{5}-\frac{1}{5}=\frac{3}{5}$ of the students

Lesson Practice 3B

1. $\frac{4}{6}$; four-sixths
2. $\frac{3}{5}$; three-fifths
3. $\frac{2}{2}$; two-halves
4. $\frac{2}{4}$; two-fourths
5. $\frac{2}{4}+\frac{2}{4}=\frac{4}{4}$
6. $\frac{3}{6}+\frac{2}{6}=\frac{5}{6}$
7. $\frac{1}{5}+\frac{3}{5}=\frac{4}{5}$
8. $\frac{5}{6}-\frac{1}{6}=\frac{4}{6}$
9. $\frac{3}{4}-\frac{1}{4}=\frac{2}{4}$
10. $\frac{2}{3}-\frac{1}{3}=\frac{1}{3}$

11. $\frac{1}{4}+\frac{1}{4}=\frac{2}{4}$ of the laps
12. $\frac{10}{10}-\frac{1}{10}=\frac{9}{10}$ of her pay
13. $\frac{2}{5}+\frac{2}{5}=\frac{4}{5}$ of the trip completed
    $\frac{5}{5}-\frac{4}{5}=\frac{1}{5}$ of the trip left
14. $\frac{5}{6}-\frac{2}{6}=\frac{3}{6}$ of the job

Lesson Practice 3C

1. $\frac{1}{3}+\frac{2}{3}=\frac{3}{3}$ three-thirds
2. $\frac{6}{6}-\frac{3}{6}=\frac{3}{6}$ three-sixths
3. $\frac{4}{5}$; four-fifths
4. $\frac{1}{4}$; one-fourth
5. done
6. $\frac{2}{5}+\frac{3}{5}=\frac{5}{5}$
7. $\frac{6}{7}+\frac{1}{7}=\frac{7}{7}$
8. $\frac{2}{6}-\frac{1}{6}=\frac{1}{6}$
9. $\frac{5}{8}-\frac{4}{8}=\frac{1}{8}$
10. $\frac{6}{10}-\frac{1}{10}=\frac{5}{10}$
11. $\frac{3}{8}+\frac{1}{8}=\frac{4}{8}$ tsp of salt
12. $\frac{3}{12}+\frac{2}{12}=\frac{5}{12}$ of the pizza eaten
13. $\frac{12}{12}-\frac{5}{12}=\frac{7}{12}$ of the pizza left
14. $\frac{9}{10}-\frac{5}{10}=\frac{4}{10}$ of a load

Systematic Review 3D

1. $\frac{1}{4}+\frac{1}{4}=\frac{2}{4}$ two-fourths
2. $\frac{5}{6}-\frac{1}{6}=\frac{4}{6}$ four-sixths
3. $\frac{1}{2}+\frac{1}{2}=\frac{2}{2}$
4. $\frac{3}{10}+\frac{6}{10}=\frac{9}{10}$
5. $\frac{1}{6}+\frac{3}{6}=\frac{4}{6}$
6. $\frac{4}{6}-\frac{1}{6}=\frac{3}{6}$
7. $\frac{3}{4}-\frac{2}{4}=\frac{1}{4}$
8. $\frac{5}{9}-\frac{3}{9}=\frac{2}{9}$
9. $10 \div 5 = 2$
   $2 \times 4 = 8$
10. $40 \div 8 = 5$
    $5 \times 5 = 25$
11. $49 \div 7 = 7$
    $7 \times 2 = 14$
12. done
13. $3+4+5=12$ ft
14. $54+39+78=171$ ft
15. $\frac{4}{9}+\frac{1}{9}=\frac{5}{9}$ of a box
16. $\frac{3}{8}+\frac{2}{8}=\frac{5}{8}$ of the problems
    $\frac{8}{8}-\frac{5}{8}=\frac{3}{8}$ of the prob. left
17. $16 \div 2 = 8$
    $8 \times 1 = 8$ people
18. $24 \div 6 = 4$
    $4 \times 5 = 20$ pages

Systematic Review 3E

1. $\frac{3}{5}$; three-fifths
2. $\frac{1}{3}$; one-third
3. $\frac{1}{5}+\frac{1}{5}=\frac{2}{5}$
4. $\frac{2}{4}+\frac{2}{4}=\frac{4}{4}$
5. $\frac{2}{8}+\frac{5}{8}=\frac{7}{8}$
6. $\frac{4}{5}-\frac{1}{5}=\frac{3}{5}$

7. $\frac{4}{6}-\frac{2}{6}=\frac{2}{6}$
8. $\frac{9}{10}-\frac{4}{10}=\frac{5}{10}$
9. $12\div4=3$
   $3x3=9$
10. $40\div5=8$
    $8x2=16$
11. $14\div7=2$
    $2x5=10$
12. $6+6+10=22$ ft
13. $28+16+28+16=88$ yd
14. $17+17+17+17=68$ in
15. $\frac{3}{8}+\frac{4}{8}=\frac{7}{8}$ of the day
16. $\frac{2}{4}+\frac{1}{4}=\frac{3}{4}$ is peanuts or almonds
    $\frac{4}{4}-\frac{3}{4}=\frac{1}{4}$ not peanuts or almonds
17. 1 flower bed:
    $6+8+12=26$ ft
    $26x2=52$ ft
18. $\frac{2}{5}+\frac{1}{5}=\frac{3}{5}$ of the apples
19. $10\div5=2$
    $2x3=$ 6 eaten
    $10-6=$ 4 left
20. $3.15+$.95+$1.25=$5.35

Systematic Review 3F

1. $\frac{3}{6}+\frac{2}{6}=\frac{5}{6}$ five-sixths
2. $\frac{4}{4}-\frac{2}{4}=\frac{2}{4}$ Two-fourths
3. $\frac{1}{3}+\frac{1}{3}=\frac{2}{3}$
4. $\frac{2}{5}+\frac{1}{5}=\frac{3}{5}$
5. $\frac{3}{11}+\frac{4}{11}=\frac{7}{11}$
6. $\frac{5}{9}-\frac{4}{9}=\frac{1}{9}$
7. $\frac{6}{7}-\frac{1}{7}=\frac{5}{7}$
8. $\frac{3}{3}-\frac{1}{3}=\frac{2}{3}$
9. $28\div7=4$
   $4x3=12$
10. $27\div3=9$
    $9x1=9$
11. $36\div4=9$
    $9x3=27$
12. $63+87+48=198$ ft
13. $35+18+35+18=106$ yd
14. $81+81+81+81=324$ in
15. $\frac{1}{5}+\frac{3}{5}=\frac{4}{5}$ of a foot
16. $10\div5=2$
    $2x4=8$ pizzas eaten
    $10-8=2$ pizzas left over
17. $\frac{3}{3}=\frac{1}{3}=\frac{2}{3}$ of a sandwich
18. $16+16+16=48$ ft
19. $17.45-$3.19=$14.26
20. $12\div6=2$
    $1x2=2$ months

Lesson Practice 4A

1. done
2. $\frac{1}{5}=\frac{2}{10}=\frac{3}{15}=\frac{4}{20}=\frac{5}{25}$
   one-fifth = two-tenths = three-fifteenths = four-twentieths = five-twenty-fifths
3. $\frac{2}{3}=\frac{4}{6}=\frac{6}{9}=\frac{8}{12}=\frac{10}{15}$
   two-thirds = four-sixths = six-ninths = eight-twelfths = ten-fifteenths
4. $\frac{3}{6}=\frac{6}{12}=\frac{9}{18}=\frac{12}{24}=\frac{15}{30}$
   three-sixths = six-twelfths = nine-eighteenths = twelve-twenty-fourths = fifteen-thirtieths
5. done
6. $\frac{3}{5}=\frac{6}{10}=\frac{9}{15}=\frac{12}{20}$

Lesson Practice 4B

1. $\frac{3}{4}=\frac{6}{8}=\frac{9}{12}=\frac{12}{16}=\frac{15}{20}$
   three-fourths = six-eighths = nine-twelfths = twelve-sixteenths = fifteen-twentieths
2. $\frac{2}{2}=\frac{4}{4}=\frac{6}{6}=\frac{8}{8}=\frac{10}{10}$
   two-halves = four-fourths = six-sixths = eight-eighths = ten-tenths
3. $\frac{4}{5}=\frac{8}{10}=\frac{12}{15}=\frac{16}{20}=\frac{20}{25}$
   four-fifths = eight-tenths = twelve-fifteenths = sixteen-twentieths = twenty-twenty-fifths
4. $\frac{1}{3}=\frac{2}{6}=\frac{3}{9}=\frac{4}{12}$
5. $\frac{2}{4}=\frac{4}{8}=\frac{6}{12}=\frac{8}{16}$
6. $\frac{5}{6}=\frac{10}{12}=\frac{15}{18}=\frac{20}{24}$
7. $\frac{3}{3}=\frac{6}{6}=\frac{9}{9}=\frac{12}{12}$
8. $\frac{1}{3}=\frac{2}{6}$

Lesson Practice 4C

1. $\frac{1}{6}=\frac{2}{12}=\frac{3}{18}=\frac{4}{24}=\frac{5}{30}$
   one-sixth = two-twelfths = three-eighteenths = four-twenty-fourths = five-thirtieths
2. $\frac{2}{5}=\frac{4}{10}=\frac{6}{15}=\frac{8}{20}=\frac{10}{25}$
   two-fifths = four-tenths = six-fifteenths = eight-twentieths = ten-twenty-fifths
3. $\frac{5}{6}=\frac{10}{12}=\frac{15}{18}=\frac{20}{24}=\frac{25}{30}$
   five-sixths = ten-twelfths = fifteen-eighteenths = twenty-twenty-fourths = twenty-five-thirtieths
4. $\frac{3}{4}=\frac{6}{8}=\frac{9}{12}=\frac{12}{16}$
5. $\frac{1}{5}=\frac{2}{10}=\frac{3}{15}=\frac{4}{20}$
6. $\frac{4}{5}=\frac{8}{10}=\frac{12}{15}=\frac{16}{20}$
7. $\frac{2}{3}=\frac{4}{6}=\frac{6}{9}=\frac{8}{12}$
8. $\frac{1}{4}=\frac{2}{8}$

Systematic Review 4D

1. $\frac{4}{4}=\frac{8}{8}=\frac{12}{12}=\frac{16}{16}=\frac{20}{20}$
   four-fourths = eight-eighths = twelve-twelfths = sixteen-sixteenths = twenty-twentieths
2. $\frac{1}{4}=\frac{2}{8}=\frac{3}{12}=\frac{4}{16}$
3. $\frac{1}{5}=\frac{2}{10}=\frac{3}{15}=\frac{4}{20}$
4. $\frac{2}{6}=\frac{4}{12}=\frac{6}{18}=\frac{8}{24}$
5. $\frac{3}{3}=\frac{6}{6}=\frac{9}{9}=\frac{12}{12}$
6. $\frac{3}{6}+\frac{2}{6}=\frac{5}{6}$
7. $\frac{5}{7}-\frac{3}{7}=\frac{2}{7}$
8. $\frac{1}{4}+\frac{1}{4}=\frac{2}{4}$
9. $36 \div 6 = 6$
   $6 \times 1 = 6$
10. $32 \div 8 = 4$
    $4 \times 5 = 20$
11. $81 \div 9 = 9$
    $9 \times 3 = 27$
12. done
13. $16 \times 11 = 176$
14. $123 \times 32 = 3{,}936$

15. $2: \frac{1}{3} \times \frac{2}{2} = \frac{2}{6}$

16. $\frac{1}{2} = \frac{4}{8}$: yes

17. $\frac{3}{4}$ of \$36

$36 \div 4 = 9$

$9 \times 3 = \$27$

18. $\frac{2}{7} + \frac{3}{7} = \frac{5}{7}$

$\frac{7}{7} - \frac{5}{7} = \frac{2}{7}$ of the chores

16. $\frac{2}{3}$ of 18:

$18 \div 3 = 6$

$6 \times 2 = 12$ students

17. $\frac{1}{2} = \frac{2}{4}$

18. $\frac{1}{4} + \frac{2}{4} = \frac{3}{4}$ of a century

19. $\frac{6}{6} - \frac{1}{6} = \frac{5}{6}$ of the bushes

20. blue : $\frac{1}{6}$ of 12

$12 \div 6 = 2; 2 \times 1 = 2$ bushes

red : $12 - 2 = 10$ bushes

Systematic Review 4E

1. $\frac{1}{3} = \frac{2}{6} = \frac{3}{9} = \frac{4}{12} = \frac{5}{15}$

one – third = two-sixths = three-ninths = four-twelfths = five-fifteenths

2. $\frac{4}{6} = \frac{8}{12} = \frac{12}{18} = \frac{16}{24}$

3. $\frac{2}{5} = \frac{4}{10} = \frac{6}{15} = \frac{8}{20}$

4. $\frac{1}{2} = \frac{2}{4} = \frac{3}{6} = \frac{4}{8}$

5. $\frac{1}{3} = \frac{2}{6} = \frac{3}{9} = \frac{4}{12}$

6. $\frac{3}{4} + \frac{1}{4} = \frac{4}{4}$

7. $\frac{6}{9} - \frac{2}{9} = \frac{4}{9}$

8. $\frac{8}{8} - \frac{5}{8} = \frac{3}{8}$

9. $49 \div 7 = 7$

$7 \times 4 = 28$

10. $45 \div 5 = 9$

$9 \times 3 = 27$

11. $12 \div 2 = 6$

$6 \times 1 = 6$

12. $22 \times 12 = 264$

13. $23 \times 13 = 299$

14. $405 \times 11 = 4{,}455$

15. $\frac{2}{3} = \frac{4}{6}$

Systematic Review 4F

1. $\frac{1}{2} = \frac{2}{4} = \frac{3}{6} = \frac{4}{8} = \frac{5}{10}$

one-half = two-fourths = three-sixths = four-eighths = five-tenths

2. $\frac{1}{6} = \frac{2}{12} = \frac{3}{18} = \frac{4}{24}$

3. $\frac{3}{5} = \frac{6}{10} = \frac{9}{15} = \frac{12}{20}$

4. $\frac{1}{7} = \frac{2}{14} = \frac{3}{21} = \frac{4}{28}$

5. $\frac{3}{8} = \frac{6}{16} = \frac{9}{24} = \frac{12}{32}$

6. $\frac{2}{3} - \frac{1}{3} = \frac{1}{3}$

7. $\frac{3}{5} + \frac{1}{5} = \frac{4}{5}$

8. $\frac{3}{10} - \frac{2}{10} = \frac{1}{10}$

9. $60 \div 10 = 6$

$6 \times 1 = 6$

10. $16 \div 8 = 2$

$2 \times 2 = 4$

11. $9 \div 3 = 3$

$3 \times 3 = 9$

12. $12 \times 11 = 132$

13. $14 \times 12 = 168$

14. $221 \times 43 = 9{,}503$

15. $\frac{1}{7}=\frac{2}{14}$

16. $\frac{2}{14}$ of 14 = 2 pounds

$14 \div 14 = 1;$

$1 \text{x} 2 = 2$

17. $11 \text{x} 13 = 143$ rolls

18. $4+4+4+4=16$ ft

19. $\frac{3}{8}+\frac{1}{8}=\frac{4}{8}$ done

$\frac{8}{8}-\frac{4}{8}=\frac{4}{8}$ left to drive

20. $\frac{4}{8}$ of 64:

$64 \div 8 = 8$

$8 \text{x} 4 = 32$ miles

Lesson Practice 5A

1. done

2. $\frac{1}{3}=\frac{2}{6}=\frac{3}{9}=\frac{4}{12}=\frac{5}{15}$

$\frac{2}{5}=\frac{4}{10}=\frac{6}{15}=\frac{8}{20}=\frac{10}{25}$

$\frac{5}{15}+\frac{6}{15}=\frac{11}{15}$

3. $\frac{1}{4}=\frac{2}{8}=\frac{3}{12}=\frac{4}{16}=\frac{5}{20}$

$\frac{1}{5}=\frac{2}{10}=\frac{3}{15}=\frac{4}{20}=\frac{5}{25}$

$\frac{5}{20}+\frac{4}{20}=\frac{9}{20}$

4. $\frac{1}{2}=\frac{2}{4}=\frac{3}{6}=\frac{4}{8}=\frac{5}{10}$

$\frac{1}{6}=\frac{2}{12}=\frac{3}{18}=\frac{4}{24}=\frac{5}{30}$

$\frac{3}{6}+\frac{1}{6}=\frac{4}{6}$

Lesson Practice 5B

1. $\frac{5}{6}=\frac{10}{12}=\frac{15}{18}=\frac{20}{24}=\frac{25}{30}$

$\frac{3}{4}=\frac{6}{8}=\frac{9}{12}=\frac{12}{16}=\frac{15}{20}$

$\frac{10}{12}-\frac{9}{12}=\frac{1}{12}$

2. $\frac{1}{2}=\frac{2}{4}=\frac{3}{6}=\frac{4}{8}=\frac{5}{10}$

$\frac{1}{3}=\frac{2}{6}=\frac{3}{9}=\frac{4}{12}=\frac{5}{15}$

$\frac{3}{6}-\frac{2}{6}=\frac{1}{6}$

3. $\frac{3}{4}=\frac{6}{8}=\frac{9}{12}=\frac{12}{16}=\frac{15}{20}$

$\frac{2}{3}=\frac{4}{6}=\frac{6}{9}=\frac{8}{12}=\frac{10}{15}$

$\frac{9}{12}-\frac{8}{12}=\frac{1}{12}$

4. $\frac{2}{4}=\frac{4}{8}=\frac{6}{12}=\frac{8}{16}=\frac{10}{20}$

$\frac{1}{5}=\frac{2}{10}=\frac{3}{15}=\frac{4}{20}=\frac{5}{25}$

$\frac{10}{20}-\frac{4}{20}=\frac{6}{20}$

Lesson Practice 5C

1. done

2. $\frac{12}{15}-\frac{5}{15}=\frac{7}{15}$

3. $\frac{10}{20}+\frac{8}{20}=\frac{18}{20}$

4. $\frac{6}{12}-\frac{4}{12}=\frac{2}{12}$

5. $\frac{6}{18}+\frac{9}{18}=\frac{15}{18}$

6. $\frac{9}{15}-\frac{5}{15}=\frac{4}{15}$

7. $\frac{10}{20}+\frac{4}{20}=\frac{14}{20}$

Systematic Review 5D

1. $\frac{3}{6}+\frac{2}{6}=\frac{5}{6}$

2. $\frac{5}{15}+\frac{6}{15}=\frac{11}{15}$

3. $\frac{12}{24}+\frac{6}{24}=\frac{18}{24}$

4. $\frac{9}{18}-\frac{6}{18}=\frac{3}{18}$

5. $\frac{12}{18}-\frac{3}{18}=\frac{9}{18}$

6. $\frac{6}{30} - \frac{5}{30} = \frac{1}{30}$
7. $\frac{2}{3} = \frac{4}{6} = \frac{6}{9} = \frac{8}{12}$
8. $\frac{4}{7} = \frac{8}{14} = \frac{12}{21} = \frac{16}{28}$
9. $24 \div 4 = 6$
   $6 \text{x} 1 = 6$
10. $25 \div 5 = 5$
    $5 \text{x} 2 = 10$
11. $63 \div 7 = 9$
    $9 \text{x} 1 = 9$
12. 30
13. 80
14. 400
15. 700
16. $\frac{5}{6}$ of 60:
    $60 \div 6 = 10$
    $10 \text{x} 5 = 50$ turkeys
17. $\frac{2}{3} + \frac{1}{6} = \frac{12}{18} + \frac{3}{18} = \frac{15}{18}$ of a pizza
18. $31 \text{x} 12 = 372$ months

11. $9 \div 9 = 1$
    $1 \text{x} 2 = 2$
12. 50
13. 60
14. 800
15. 100
16. $\frac{1}{3} + \frac{1}{2} + \frac{2}{6} + \frac{3}{6} = \frac{5}{6}$ of a mile
17. $\frac{2}{5} + \frac{1}{4} = \frac{8}{20} + \frac{5}{20} = \frac{13}{20}$
    of the children
18. $\frac{2}{5}$ of 20:
    $20 \div 5 = 4$
    $4 \text{x} 2 = 8$ children bought pears
    $\frac{1}{4}$ of 20:
    $20 \div 4 = 5$
    $5 \text{x} 1 = 5$ children bought apples
19. $221 \text{x} 4 = 884$ dirty paws
20. $8 + 13 + 19 = 40$ ft

Systematic Review 5E

1. $\frac{6}{30} + \frac{5}{30} = \frac{11}{30}$
2. $\frac{5}{15} + \frac{9}{15} = \frac{14}{15}$
3. $\frac{5}{20} + \frac{4}{20} = \frac{9}{20}$
4. $\frac{25}{30} - \frac{6}{30} = \frac{19}{30}$
5. $\frac{16}{20} - \frac{15}{20} = \frac{1}{20}$
6. $\frac{6}{18} - \frac{3}{18} = \frac{3}{18}$
7. $\frac{1}{2} = \frac{2}{4} = \frac{3}{6} = \frac{4}{8}$
8. $\frac{3}{6} = \frac{6}{12} = \frac{9}{18} = \frac{12}{24}$
9. $18 \div 3 = 6$
   $6 \text{x} 1 = 6$
10. $42 \div 7 = 6$
    $6 \text{x} 3 = 18$

Systematic Review 5F

1. $\frac{12}{24} + \frac{8}{24} = \frac{20}{24}$
2. $\frac{5}{20} + \frac{12}{20} = \frac{17}{20}$
3. $\frac{6}{18} + \frac{6}{18} = \frac{12}{18}$
4. $\frac{3}{6} - \frac{2}{6} = \frac{1}{6}$
5. $\frac{8}{20} - \frac{5}{20} = \frac{3}{20}$
6. $\frac{15}{18} - \frac{6}{18} = \frac{9}{18}$
7. $\frac{3}{4} = \frac{6}{8} = \frac{9}{12} = \frac{12}{16}$
8. $\frac{5}{9} = \frac{10}{18} = \frac{15}{27} = \frac{20}{36}$
9. $25 \div 5 = 5$
   $5 \text{x} 2 = 10$
10. $30 \div 3 = 10$
    $10 \text{x} 2 = 20$
11. $56 \div 8 = 7$
    $7 \text{x} 1 = 7$

12. 60
13. 10
14. 100
15. 400
16. $\frac{3}{4} - \frac{1}{2} = \frac{6}{8} - \frac{4}{8} = \frac{2}{8}$ of a yard
17. $\frac{3}{8}$ of 80:
    $80 \div 8 = 10$
    10x3 = 30 people with flags
18. $15.34 + $19.99 + $16.50 = $51.83
19. 50 + 30 + 50 + 30 = 160 ft
20. 12x24 = 288 people

Lesson Practice 6A

1. done
2. $\frac{5}{10} + \frac{2}{10} = \frac{7}{10}$
3. $\frac{4}{12} + \frac{6}{12} = \frac{10}{12}$
4. $\frac{12}{18} - \frac{6}{18} = \frac{6}{18}$
5. $\frac{12}{30} - \frac{5}{30} = \frac{7}{30}$
6. $\frac{5}{20} - \frac{4}{20} = \frac{1}{20}$

Lesson Practice 6B

1. done
2. $\frac{5}{20} + \frac{12}{20} = \frac{17}{20}$
3. $\frac{3}{18} + \frac{6}{18} = \frac{9}{18}$
4. $\frac{8}{12} + \frac{3}{12} = \frac{11}{12}$
5. $\frac{20}{30} - \frac{6}{30} = \frac{14}{30}$
6. $\frac{4}{8} - \frac{2}{8} = \frac{2}{8}$
7. $\frac{12}{15} - \frac{10}{15} = \frac{2}{15}$
8. $\frac{6}{12} - \frac{4}{12} = \frac{2}{12}$
9. $\frac{3}{4} + \frac{1}{6} = \frac{18}{24} + \frac{4}{24} = \frac{22}{24}$
   of the casserole
10. $\frac{3}{4} - \frac{1}{2} = \frac{6}{8} - \frac{4}{8} = \frac{2}{8}$ in

Lesson Practice 6C

1. $\frac{4}{12} + \frac{3}{12} = \frac{7}{12}$
2. $\frac{5}{35} + \frac{14}{35} = \frac{19}{35}$
3. $\frac{6}{16} + \frac{8}{16} = \frac{14}{16}$
4. $\frac{4}{24} + \frac{18}{24} = \frac{22}{24}$
5. $\frac{25}{30} - \frac{18}{30} = \frac{7}{30}$
6. $\frac{16}{20} - \frac{5}{20} = \frac{11}{20}$
7. $\frac{5}{15} - \frac{3}{15} = \frac{2}{15}$
8. $\frac{14}{18} - \frac{9}{18} = \frac{5}{18}$
9. $\frac{1}{2} + \frac{1}{4} = \frac{4}{8} + \frac{2}{8} = \frac{6}{8}$
   of her books
10. $\frac{2}{5} - \frac{1}{3} = \frac{6}{15} - \frac{5}{15} = \frac{1}{15}$
    of a bucket

Systematic Review 6D

1. $\frac{15}{20} + \frac{4}{20} = \frac{19}{20}$
2. $\frac{6}{24} + \frac{8}{24} = \frac{14}{24}$

3. $\frac{12}{18}+\frac{3}{18}=\frac{15}{18}$
4. $\frac{6}{24}-\frac{4}{24}=\frac{2}{24}$
5. $\frac{10}{30}-\frac{6}{30}=\frac{4}{30}$
6. $\frac{32}{72}-\frac{9}{72}=\frac{23}{72}$
7. $\frac{1}{5}=\frac{2}{10}=\frac{3}{15}=\frac{4}{20}$
8. $\frac{4}{5}=\frac{8}{10}=\frac{12}{15}=\frac{16}{20}$
9. $12 \div 3 = 4$
   $4 \text{x} 1 = 4$
10. $16 \div 8 = 2$
    $2 \text{x} 7 = 14$
11. $81 \div 9 = 9$
    $9 \text{x} 4 = 36$
12. $(20)\text{x}(40) = (800)$
    $23 \text{x} 36 = 828$
13. $(80)\text{x}(30) = (2{,}400)$
    $78 \text{x} 34 = 2{,}652$
14. $(70)\text{x}(20) = (1{,}400)$
    $65 \text{x} 15 = 975$
15. 15x24 = 360 tomatoes
16. $\frac{2}{3}$ of \$9 = \$6 on hand
    \$9 - \$6 = \$3 needed
17. $\frac{1}{5}+\frac{1}{6}=\frac{6}{30}+\frac{5}{30}=\frac{11}{30}$ of the troops
18. $\frac{1}{2}+\frac{2}{5}=\frac{5}{10}+\frac{4}{10}=\frac{9}{10}$ in

Systematic Review 6E

1. $\frac{6}{30}+\frac{15}{30}=\frac{21}{30}$
2. $\frac{3}{21}+\frac{14}{21}=\frac{17}{21}$
3. $\frac{5}{15}+\frac{3}{15}=\frac{8}{15}$
4. $\frac{15}{18}-\frac{12}{18}=\frac{3}{18}$
5. $\frac{4}{12}-\frac{3}{12}=\frac{1}{12}$
6. $\frac{9}{18}-\frac{2}{18}=\frac{7}{18}$
7. $\frac{3}{7}=\frac{6}{14}=\frac{9}{21}=\frac{12}{28}$
8. $\frac{9}{11}=\frac{18}{22}=\frac{27}{33}=\frac{36}{44}$
9. $35 \div 7 = 5$
   $5 \text{x} 1 = 5$
10. $50 \div 5 = 10$
    $10 \text{x} 4 = 40$
11. $36 \div 4 = 9$
    $9 \text{x} 3 = 27$
12. $(50)\text{x}(20) = (1{,}000)$
    $45 \text{x} 24 = 1{,}080$
13. $(70)\text{x}(20) = (1{,}400)$
    $67 \text{x} 18 = 1{,}206$
14. $(30)\text{x}(40) = (1{,}200)$
    $32 \text{x} 39 = 1{,}248$
15. 15x\$33 = \$495
16. $80 \div 10 = 8$
    8x9 = 72 posts
17. $\frac{3}{6}+\frac{1}{6}=\frac{4}{6}$ gone
    $\frac{6}{6}-\frac{4}{6}=\frac{2}{6}$ left
18. $\frac{1}{6}-\frac{1}{12}=\frac{12}{72}-\frac{6}{72}=\frac{6}{72}$ ft
19. 500
20. 6+6+6+6 = 24 yds

Systematic Review 6F

1. $\frac{20}{30}+\frac{6}{30}=\frac{26}{30}$
2. $\frac{6}{12}+\frac{4}{12}=\frac{10}{12}$
3. $\frac{22}{77}+\frac{21}{77}=\frac{43}{77}$
4. $\frac{5}{10}-\frac{2}{10}=\frac{3}{10}$
5. $\frac{8}{12}-\frac{3}{12}=\frac{5}{12}$
6. $\frac{18}{72}-\frac{16}{72}=\frac{2}{72}$
7. $\frac{5}{6}=\frac{10}{12}=\frac{15}{18}=\frac{20}{24}$
8. $\frac{1}{10}=\frac{2}{20}=\frac{3}{30}=\frac{4}{40}$

9. $28 \div 7 = 4;$
   $4x3 = 12$
10. $54 \div 6 = 9;\ 9x1 = 9$
11. $8 \div 8 = 1;\ 1x4 = 4$
12. $(70)x(90) = (6{,}300)$
    $73x89 = 6{,}497$
13. $(30)x(90) = (2{,}700)$
    $26x91 = 2{,}366$
14. $(50)x(10) = (500)$
    $47x11 = 517$
15. $18x12 = 216$ eggs
16. $312x3 = 936$ mi
17. $\frac{4}{9} + \frac{3}{6} = \frac{24}{54} + \frac{27}{54} = \frac{51}{54}$ of a loaf
18. $60 \div 6 = 10$
    $5x10 = 50$ min
19. 300
20. $13+18+13+18 = 62$ in
    $2x24 = 48$ in
    62 in > 48 in; no

Lesson Practice 7A

1. done
2. $\frac{9}{15} < \frac{10}{15}$ so $\frac{3}{5} < \frac{2}{3}$
3. done
4. $\frac{18}{30} < \frac{20}{30}$ so $\frac{3}{5} < \frac{4}{6}$
5. $\frac{8}{12} < \frac{9}{12}$ so $\frac{2}{3} < \frac{3}{4}$
6. $\frac{6}{15} > \frac{5}{15}$ so $\frac{2}{5} > \frac{1}{3}$

Lesson Practice 7B

1. done
2. $\frac{12}{18} < \frac{15}{18}$ so $\frac{2}{3} < \frac{5}{6}$
   less than
3. $\frac{12}{18} > \frac{9}{18}$ so $\frac{2}{3} > \frac{3}{6}$
   greater than
4. $\frac{5}{10} > \frac{4}{10}$ so $\frac{1}{2} > \frac{2}{5}$
   greater than
5. $\frac{6}{18} = \frac{6}{18}$ so $\frac{1}{3} = \frac{2}{6}$
   equal
6. $\frac{10}{20} > \frac{4}{20}$ so $\frac{2}{4} > \frac{1}{5}$
   greater than
7. $\frac{5}{10} > \frac{4}{10}$ so $\frac{1}{2} > \frac{2}{5}$
   Trisha got more votes.
8. $\frac{6}{18} < \frac{12}{18}$ so $\frac{2}{6} < \frac{2}{3}$
   Donald ran further.

Lesson Practice 7C

1. $\frac{24}{30} > \frac{20}{30}$ so $\frac{4}{5} > \frac{4}{6}$
   greater than
2. $\frac{8}{12} < \frac{12}{12}$ so $\frac{4}{6} < \frac{2}{2}$
   less than
3. $\frac{21}{56} < \frac{32}{56}$ so $\frac{3}{8} < \frac{4}{7}$
4. $\frac{6}{27} < \frac{9}{27}$ so $\frac{2}{9} < \frac{1}{3}$
   less than
5. $\frac{18}{24} < \frac{20}{24}$
6. $\frac{12}{24} = \frac{12}{24}$
7. $\frac{10}{20} > \frac{6}{20}$
8. $\frac{28}{35} < \frac{30}{35}$
9. $\frac{6}{24} > \frac{4}{24}$ Shirley ate more pizza.
10. $\frac{36}{60} > \frac{35}{60}$ On the east side is larger.

Systematic Review 7D

1. $\frac{6}{18} < \frac{9}{18}$

2. $\frac{10}{16} > \frac{8}{16}$
3. $\frac{12}{48} = \frac{12}{48}$
4. $\frac{12}{24} + \frac{4}{24} = \frac{16}{24}$
5. $\frac{48}{80} - \frac{30}{80} = \frac{18}{80}$
6. $\frac{14}{63} + \frac{45}{63} = \frac{59}{63}$
7. $\frac{6}{8} = \frac{12}{16} = \frac{18}{24} = \frac{24}{32}$
8. $6 \div 2 = 3;$
   $3x1 = 3$
9. $42 \div 6 = 7;$
   $7x3 = 21$
10. $24 \div 8 = 3;$
    $3x3 = 9$
11. done
12. $23 \div 5 = 4\frac{3}{5}$
13. $59 \div 7 = 8\frac{3}{7}$
14. $17 \div 4 = 4\frac{1}{4}$ yd
15. $\frac{16}{56} < \frac{35}{56}$ Penny did the most.
    $\frac{16}{56} + \frac{35}{56} = \frac{51}{56}$ done
    $\frac{56}{56} - \frac{51}{56} = \frac{5}{56}$ left
16. $\frac{5}{56}$ of 56:
    $56 \div 56 = 1$; 1x5=5 chores
17. $\frac{3}{12} + \frac{4}{12} = \frac{7}{12}$ of a cup
18. $\frac{21}{24} - \frac{8}{24} = \frac{13}{24}$ in

Systematic Review 7E

1. $\frac{9}{15} > \frac{5}{15}$
2. $\frac{12}{18} > \frac{3}{18}$
3. $\frac{108}{120} > \frac{70}{120}$
4. $\frac{5}{10} + \frac{4}{10} = \frac{9}{10}$
5. $\frac{6}{12} - \frac{4}{12} = \frac{2}{12}$
6. $\frac{15}{40} + \frac{24}{40} = \frac{39}{40}$
7. $\frac{1}{10} = \frac{2}{20} = \frac{3}{30} = \frac{4}{40}$
8. $32 \div 8 = 4$; 4x7=28
9. $21 \div 7 = 3$; 3x2=6
10. $20 \div 4 = 5$; 5x3=15
11. $32 \div 6 = 5\frac{2}{6}$
12. $19 \div 8 = 2\frac{3}{8}$
13. $48 \div 5 = 9\frac{3}{5}$
14. (20)x(20) = (400)
    21x16 = 336
15. (30)x(30) = (900)
    34x29 = 986
16. (80)x(10) = (800)
    75x12 = 900
17. $\frac{7}{42} + \frac{6}{42} = \frac{13}{42}$ of the cars
18. $28 \div 7 = 4$
    4x4 = 16 games
19. $8 + 10 + 8 + 10 = 36$
    $\frac{1}{4}$ of 36 = 9 yd
20. $\frac{8}{16} > \frac{6}{16}$ so $\frac{1}{2} > \frac{3}{8}$

Systematic Review 7F

1. $\frac{60}{120} = \frac{60}{120}$
2. $\frac{10}{35} < \frac{21}{35}$
3. $\frac{3}{6} < \frac{4}{6}$
4. $\frac{10}{15} + \frac{3}{15} = \frac{13}{15}$
5. $\frac{16}{24} - \frac{6}{24} = \frac{10}{24}$
6. $\frac{45}{54} + \frac{6}{54} = \frac{51}{54}$

7. $\frac{3}{4} = \frac{6}{8} = \frac{9}{12} = \frac{12}{16}$
8. $10 \div 5 = 2;\ 2x3 = 6$
9. $12 \div 4 = 3;\ 3x1 = 3$
10. $24 \div 6 = 4;\ 4x4 = 16$
11. $13 \div 3 = 4\frac{1}{3}$
12. $39 \div 4 = 9\frac{3}{4}$
13. $58 \div 9 = 6\frac{4}{9}$
14. $(60)x(50) = (3{,}000)$
    $65x51 = 3{,}264$
15. $(50)x(20) = (1{,}000)$
    $45x19 = 855$
16. $(80)x(40) = (3{,}200)$
    $82x37 = 3{,}034$
17. $8 + 9 + 10 = 27$ ft
18. $\frac{25}{30} > \frac{24}{30}$ so $\frac{5}{6} > \frac{4}{5}$
19. $30 \div 6 = 5$
    $5x5 = 25$ questions for Kiley
    $30 \div 5 = 6$
    $6x4 = 24$ questions for Casey
    $25 > 24$; yes
20. $\frac{20}{32} < \frac{24}{32}$ so $\frac{5}{8} < \frac{3}{4}$

Lesson Practice 8A

1. done
2. $\frac{6}{30} + \frac{5}{30} = \frac{11}{30}$
   $\frac{11}{30} + \frac{1}{2} = \frac{22}{60} + \frac{30}{60} = \frac{52}{60}$
3. $\frac{3}{24} + \frac{8}{24} = \frac{11}{24}$
   $\frac{11}{24} + \frac{1}{6} = \frac{66}{144} + \frac{24}{144} = \frac{90}{144}$
4. $\frac{6}{24} + \frac{20}{24} = \frac{26}{24}$
   $\frac{26}{24} + \frac{1}{3} = \frac{78}{72} + \frac{24}{72} = \frac{102}{72} = 1\frac{30}{72}$
   Note: at this point the final step is optional when working with fractions larger than one.
5. done
6. done
7. $\frac{3x6x10}{7x6x10} + \frac{1x7x10}{6x7x10} + \frac{3x7x6}{10x7x6} =$
   $\frac{180}{420} + \frac{70}{420} + \frac{126}{420} = \frac{376}{420}$
8. $\frac{4x2}{5x2} + \frac{3}{10} + \frac{1x5}{2x5} =$
   $\frac{8}{10} + \frac{3}{10} + \frac{5}{10} = \frac{16}{10} = 1\frac{6}{10}$
9. $\frac{5}{8} + \frac{7}{8} + \frac{4}{8} = \frac{16}{8} = 2$ ft
10. $\frac{5}{40} + \frac{8}{40} = \frac{13}{40}$
    $\frac{13}{40} + \frac{1}{4} = \frac{52}{160} + \frac{40}{160} = \frac{92}{160}$
    or $\frac{13}{40} + \frac{10}{40} = \frac{23}{40}$ of a tank

The student may begin to recognize shortcuts that yield fractions that look different than the answers given. As long as the answer is equivalent to the given answer it is correct. For example, 2/5 is the same as 4/10.

Lesson Practice 8B

1. $\frac{5}{10} + \frac{4}{10} = \frac{9}{10}$
   $\frac{9}{10} + \frac{5}{6} = \frac{54}{60} + \frac{50}{60} = \frac{104}{60} = 1\frac{44}{60}$
2. $\frac{10}{35} + \frac{7}{35} = \frac{17}{35}$
   $\frac{17}{35} + \frac{2}{3} = \frac{51}{105} + \frac{70}{105} = \frac{121}{105} = 1\frac{16}{105}$
3. $\frac{6}{16} + \frac{8}{16} = \frac{14}{16}$
   $\frac{14}{16} + \frac{2}{5} = \frac{70}{80} + \frac{32}{80} = \frac{102}{80} = 1\frac{22}{80}$
4. $\frac{9}{63} + \frac{14}{63} = \frac{23}{63}$
   $\frac{23}{63} + \frac{1}{3} = \frac{69}{189} + \frac{63}{189} = \frac{132}{189}$
5. $\frac{5x2x3}{7x2x3} + \frac{1x7x3}{2x7x3} + \frac{1x7x2}{3x7x2} =$
   $\frac{30}{42} + \frac{21}{42} + \frac{14}{42} = \frac{65}{42} = 1\frac{23}{42}$
6. $\frac{4}{8} + \frac{6}{8} + \frac{5}{8} = \frac{15}{8} = 1\frac{7}{8}$

7. $\frac{2x4x6}{5x4x6}+\frac{1x5x6}{4x5x6}+\frac{1x5x4}{6x5x4}=$
   $\frac{48}{120}+\frac{30}{120}+\frac{20}{120}=\frac{98}{120}$
8. $\frac{4}{9}+\frac{3}{9}+\frac{2}{9}=\frac{9}{9}=1$
9. $\frac{1}{10}+\frac{4}{10}+\frac{5}{10}=\frac{10}{10}=1$ mile
10. $\frac{5}{6}+\frac{3}{6}+\frac{4}{6}=\frac{12}{6}=2$ oranges

Lesson Practice 8C

1. $\frac{6}{10}+\frac{5}{10}=\frac{11}{10}$
   $\frac{11}{10}+\frac{1}{3}=\frac{33}{30}+\frac{10}{30}=\frac{43}{30}=1\frac{13}{30}$
2. $\frac{15}{18}+\frac{6}{18}=\frac{21}{18}$
   $\frac{21}{18}+\frac{2}{5}=\frac{105}{90}+\frac{36}{90}=\frac{141}{90}=1\frac{51}{90}$
   or $\frac{47}{30}=1\frac{7}{30}$
3. $\frac{4}{12}+\frac{9}{12}=\frac{13}{12}$
   $\frac{13}{12}+\frac{5}{6}=\frac{78}{72}+\frac{60}{72}=\frac{138}{72}=1\frac{66}{72}$
   or $\frac{46}{24}=1\frac{22}{24}$
4. $\frac{6}{18}+\frac{6}{18}=\frac{12}{18}$
   $\frac{12}{18}+\frac{1}{2}=\frac{24}{36}+\frac{18}{36}=\frac{42}{36}=1\frac{6}{36}$ or $\frac{7}{6}=1\frac{1}{6}$
5. $\frac{1}{5}+\frac{1}{5}=\frac{2}{5}$
   $\frac{2}{5}+\frac{3}{10}=\frac{20}{50}+\frac{15}{50}=\frac{35}{50}$ or $\frac{7}{10}$
6. $\frac{8}{20}+\frac{5}{20}=\frac{13}{20}$
   $\frac{13}{20}+\frac{1}{6}=\frac{78}{120}+\frac{20}{120}=\frac{98}{120}$
7. $\frac{12}{14}+\frac{3}{14}+\frac{7}{14}=\frac{22}{14}=1\frac{8}{14}$ or $\frac{308}{196}=1\frac{112}{196}$
8. $\frac{28}{32}+\frac{8}{32}=\frac{36}{32}$
   $\frac{36}{32}+\frac{1}{3}=\frac{108}{96}+\frac{32}{96}=\frac{140}{96}=1\frac{44}{96}$
9. $\frac{1}{4}+\frac{1}{4}=\frac{2}{4}$
   $\frac{2}{4}+\frac{1}{8}=\frac{16}{32}+\frac{4}{32}=\frac{20}{32}$ or $\frac{5}{8}$ of the job
10. $\frac{4}{8}+\frac{2}{8}+\frac{1}{8}=\frac{7}{8}$ or $\frac{56}{64}$ of her money

Systematic Review 8D

1. $\frac{8}{20}+\frac{5}{20}=\frac{13}{20}$
   $\frac{13}{20}+\frac{1}{2}=\frac{26}{40}+\frac{20}{40}=\frac{46}{40}=1\frac{6}{40}$
   or $\frac{23}{20}=1\frac{3}{20}$
2. $\frac{18}{20}+\frac{10}{20}=\frac{28}{20}$
   $\frac{28}{20}+\frac{3}{4}=\frac{112}{80}+\frac{60}{80}=\frac{172}{80}=2\frac{12}{80}$
   or $\frac{86}{40}=2\frac{6}{40}$
3. $\frac{1}{6}+\frac{4}{6}+\frac{3}{6}=\frac{8}{6}=1\frac{2}{6}$ or $\frac{48}{36}=1\frac{12}{36}$
4. $\frac{6}{24}>\frac{4}{24}$
5. $\frac{9}{15}>\frac{5}{15}$
6. $\frac{12}{24}=\frac{12}{24}$
7. $\frac{2}{3}=\frac{4}{6}=\frac{6}{9}=\frac{8}{12}$
8. $\frac{1}{6}=\frac{2}{12}=\frac{3}{18}=\frac{4}{24}$
9. 73x62 = 4,526
10. 54x28 = 1,512
11. 91x49 = 4,459
12. done
13. $379\div 6=63\frac{1}{6}$
14. $503\div 2=251\frac{1}{2}$
15. $\frac{8}{16}+\frac{2}{16}+\frac{1}{16}=\frac{11}{16}$
    $\frac{11}{16}<\frac{16}{16}$ Something else was used.
16. $35\div 4=8$ r.3; 8 groups, 3 left over

17. $\frac{3}{4}$ of 8 = 6 books
18. $\frac{7}{42} < \frac{12}{42}$ so $\frac{1}{6} < \frac{2}{7}$
    Clara received more.
20. $\frac{3}{6} > \frac{2}{6}$ Jeremy grew more.
    $\frac{3}{6} - \frac{2}{6} = \frac{1}{6}$ ft

Systematic Review 8E

1. $\frac{20}{32} + \frac{8}{32} = \frac{28}{32}$
   $\frac{28}{32} + \frac{1}{2} = \frac{56}{64} + \frac{32}{64} = \frac{88}{64} = 1\frac{24}{64}$
   or $\frac{11}{8} = 1\frac{3}{8}$
2. $\frac{4}{20} + \frac{10}{20} = \frac{14}{20}$
   $\frac{14}{20} + \frac{5}{6} = \frac{84}{120} + \frac{100}{120} = \frac{184}{120} = 1\frac{64}{120}$
3. $\frac{3}{6} + \frac{2}{6} = \frac{5}{6}$
   $\frac{5}{6} + \frac{3}{6} = \frac{8}{6} = 1\frac{2}{6}$
4. $\frac{12}{18} > \frac{6}{18}$
5. $\frac{20}{32} > \frac{8}{32}$
6. $\frac{14}{35} < \frac{15}{35}$
7. $\frac{1}{8} = \frac{2}{16} = \frac{3}{24} = \frac{4}{32}$
8. $\frac{1}{2} = \frac{2}{4} = \frac{3}{6} = \frac{4}{8}$
9. 35x22 = 770
10. 47x84 = 3,948
11. 63x19 = 1,197
12. 198 ÷ 3 = 66
13. $809 \div 8 = 101\frac{1}{8}$
14. 472 ÷ 4 = 118
15. $\frac{3}{6} + \frac{2}{6} + \frac{1}{6} = \frac{6}{6} = 1$ pizza
16. $\frac{1}{2} = \frac{2}{4}$ of a pie
17. 45 ÷ 4 = 11 r.1; 11 flashlights, 1 battery left
18. 122x2 = 244 hands
19. 85 ÷ 5 = 17; 17x2=34 dimes

Systematic Review 8F

1. $\frac{28}{40} + \frac{30}{40} = \frac{58}{40}$
   $\frac{58}{40} + \frac{1}{3} = \frac{174}{120} + \frac{40}{120} = \frac{214}{120} = 1\frac{94}{120}$
2. $\frac{18}{21} + \frac{7}{21} = \frac{25}{21}$
   $\frac{25}{21} + \frac{5}{6} = \frac{150}{126} + \frac{105}{126} = \frac{255}{126} = 2\frac{3}{126}$
3. $\frac{14}{16} + \frac{5}{16} + \frac{8}{16} = \frac{27}{16} = 1\frac{11}{16}$
4. $\frac{8}{18} < \frac{9}{18}$
5. $\frac{35}{42} < \frac{36}{42}$
6. $\frac{16}{24} > \frac{15}{24}$
7. $\frac{4}{7} = \frac{8}{14} = \frac{12}{21} = \frac{16}{28}$
8. $\frac{1}{3} = \frac{2}{6} = \frac{3}{9} = \frac{4}{12}$
9. 32x55 = 1,760
10. 76x41 = 3,116
11. 29x17 = 493
12. $361 \div 7 = 51\frac{4}{7}$
13. $734 \div 9 = 81\frac{5}{9}$
14. $108 \div 5 = 21\frac{3}{5}$
15. $\frac{2}{10} + \frac{1}{10} + \frac{5}{10} = \frac{8}{10}$
    $\frac{10}{10} - \frac{8}{10} = \frac{2}{10}$ of his allowance
16. 90
17. 100
18. 5 + 3 + 5 + 3 = 16 mi
19. 28 ÷ 7 = 4; 4x1 = 4 days

20. $25 \div 4 = 6\frac{1}{4}$ chocolates

Remember, equivalent answers to addition problems are correct.

Lesson Practice 9A

1. done
2. $\frac{3x3}{5x4} = \frac{9}{20}$
3. $\frac{1x1}{3x2} = \frac{1}{6}$
4. $\frac{4x2}{6x5} = \frac{8}{30}$
5. $\frac{2x1}{3x4} = \frac{2}{12}$
6. $\frac{1x5}{5x6} = \frac{5}{30}$
7. $\frac{4x2}{5x6} = \frac{8}{30}$
8. $\frac{1x1}{6x3} = \frac{1}{18}$
9. $\frac{1x2}{2x6} = \frac{2}{12}$
10. $\frac{3}{5} x \frac{2}{5} = \frac{6}{25}$
11. $\frac{3}{6} x \frac{2}{3} = \frac{6}{18}$
12. $\frac{2}{4} x \frac{1}{5} = \frac{2}{20}$
13. $\frac{3}{4} x \frac{4}{6} = \frac{12}{24}$
14. $\frac{2}{5} x \frac{1}{2} = \frac{2}{10}$
15. $\frac{4}{6} x \frac{1}{3} = \frac{4}{18}$
16. $\frac{1}{3} x \frac{1}{4} = \frac{1}{12}$ of a cup
17. $\frac{1}{2} x \frac{1}{5} = \frac{1}{10}$ of the group
18. $\frac{1}{3} x \frac{1}{5} = \frac{1}{15}$ of a pie

Lesson Practice 9B

1. $\frac{2x3}{6x5} = \frac{6}{30}$
2. $\frac{2x1}{3x2} = \frac{2}{6}$
3. $\frac{3x2}{8x4} = \frac{6}{32}$
4. $\frac{1x1}{4x5} = \frac{1}{20}$
5. $\frac{1x2}{6x3} = \frac{2}{18}$
6. $\frac{1x4}{3x5} = \frac{4}{15}$
7. $\frac{4x3}{5x6} = \frac{12}{30}$
8. $\frac{1x5}{2x6} = \frac{5}{12}$
9. $\frac{3x3}{4x4} = \frac{9}{16}$
10. $\frac{2}{5} x \frac{4}{6} = \frac{8}{30}$
11. $\frac{1}{7} x \frac{1}{2} = \frac{1}{14}$
12. $\frac{3}{5} x \frac{5}{6} = \frac{15}{30}$
13. $\frac{3}{10} x \frac{1}{4} = \frac{3}{40}$
14. $\frac{1}{4} x \frac{1}{2} = \frac{1}{8}$
15. $\frac{7}{8} x \frac{1}{9} = \frac{7}{72}$
16. $\frac{3}{4} x \frac{1}{4} = \frac{3}{16}$ of the class
17. $\frac{2}{3} x \frac{3}{5} = \frac{6}{15}$ of a bushel
18. $\frac{1}{5} x \frac{1}{5} = \frac{1}{25}$ of the games

Lesson Practice 9C

1. $\frac{3x2}{6x5} = \frac{6}{30}$
2. $\frac{2x2}{9x3} = \frac{4}{27}$

3. $\frac{2\text{x}1}{6\text{x}3} = \frac{2}{18}$
4. $\frac{3\text{x}4}{4\text{x}5} = \frac{12}{20}$
5. $\frac{1\text{x}3}{7\text{x}4} = \frac{3}{28}$
6. $\frac{1\text{x}1}{6\text{x}6} = \frac{1}{36}$
7. $\frac{4\text{x}1}{8\text{x}3} = \frac{4}{24}$
8. $\frac{3\text{x}3}{5\text{x}4} = \frac{9}{20}$
9. $\frac{4\text{x}2}{6\text{x}4} = \frac{8}{24}$
10. $\frac{1\text{x}2}{6\text{x}5} = \frac{2}{30}$
11. $\frac{5}{9} \text{x} \frac{1}{4} = \frac{5}{36}$
12. $\frac{1\text{x}1}{10\text{x}10} = \frac{1}{100}$
13. $\frac{2\text{x}3}{3\text{x}6} = \frac{6}{18}$
14. $\frac{3\text{x}1}{5\text{x}2} = \frac{3}{10}$
15. $\frac{1\text{x}3}{8\text{x}5} = \frac{3}{40}$
16. $\frac{4}{5} \text{x} \frac{2}{4} = \frac{8}{20}$ mi
17. $\frac{2}{3} \text{x} \frac{1}{3} = \frac{2}{9}$ of a cup
18. $\frac{7}{8} \text{x} \frac{1}{7} = \frac{7}{56}$ of a pie

Systematic Review 9D

1. $\frac{2}{5} \times \frac{1}{5} = \frac{2}{25}$
2. $\frac{5}{6} \times \frac{1}{4} = \frac{5}{24}$
3. $\frac{5}{9} \times \frac{1}{2} = \frac{5}{18}$
4. $\frac{3}{21} + \frac{7}{21} = \frac{10}{21}$
5. $\frac{14}{21} - \frac{3}{21} = \frac{11}{21}$
6. $\frac{5}{10} + \frac{4}{10} + \frac{7}{10} = \frac{16}{10}$ or $1\frac{6}{10}$
7. $\frac{28}{42} > \frac{6}{42}$
8. $\frac{16}{24} > \frac{15}{24}$
9. $\frac{11}{99} < \frac{18}{99}$
10. $\frac{2}{7} = \frac{4}{14} = \frac{6}{21} = \frac{8}{28}$
11. $\frac{1}{9} = \frac{2}{18} = \frac{3}{27} = \frac{4}{36}$
12. done
13. $(200) \times (60) = (12{,}000)$

    $179 \times 57 = 10{,}203$
14. $(900) \times (10) = (9{,}000)$

    $902 \times 11 = 9{,}922$
15. $\frac{1}{3} \times \frac{2}{3} = \frac{2}{9}$ of a cup
16. $365 \times 25 = 9{,}125$ days
17. $\frac{3}{6} + \frac{2}{6} = \frac{5}{6}$

    $\frac{5}{6} + \frac{1}{4} = \frac{20}{24} + \frac{6}{24} = \frac{26}{24}$

    of $1\frac{2}{24}$ cups
18. $36 \div 4 = 9$

    $9 \times 3 = 27$ hawks

Systematic Review 9E

1. $\frac{1}{2} \text{x} \frac{2}{3} = \frac{2}{6}$
2. $\frac{7}{8} \text{x} \frac{2}{5} = \frac{14}{40}$
3. $\frac{1}{3} \text{x} \frac{3}{5} = \frac{3}{15}$
4. $\frac{27}{36} + \frac{4}{36} = \frac{31}{36}$
5. $\frac{10}{15} - \frac{6}{15} = \frac{4}{15}$
6. $\frac{3}{8} + \frac{5}{8} = \frac{8}{8} = 1$

    $1 + \frac{2}{8} = 1\frac{2}{8}$ or $1\frac{1}{4}$
7. $\frac{9}{27} = \frac{9}{27}$
8. $\frac{16}{40} > \frac{15}{40}$
9. $\frac{50}{70} > \frac{49}{70}$

11. $(300)\text{x}(40) = (12{,}000)$
    $254\text{x}35 = 8{,}890$
12. $(600)\text{x}(30) = (18{,}000)$
    $563\text{x}26 = 14{,}638$
13. $107 \div 6 = 17\frac{5}{6}$
14. $395 \div 8 = 49\frac{3}{8}$
15. $459 \div 2 = 229\frac{1}{2}$
16. $\frac{1}{2} \text{x} \frac{4}{5} = \frac{4}{10}$ of the chores
17. $\frac{5}{30} + \frac{6}{30} = \frac{11}{30}$
    $\frac{11}{30} + \frac{1}{4} = \frac{44}{120} + \frac{30}{120} = \frac{74}{120}$ of the job
18. $250\text{x}51 = 12{,}750$ in
19. $12 + 17 + 28 = 57$ in
20. $235 \div 5 = 47;\ 47\text{x}3 = 141$ jelly beans

Systematic Review 9F

1. $\frac{2}{4} \text{x} \frac{2}{5} = \frac{4}{20}$
2. $\frac{1}{3} \text{x} \frac{2}{6} = \frac{2}{18}$
3. $\frac{1}{2} \text{x} \frac{4}{9} = \frac{4}{18}$
4. $\frac{10}{15} + \frac{3}{15} = \frac{13}{15}$
5. $\frac{16}{40} - \frac{15}{40} = \frac{1}{40}$
6. $\frac{5}{20} + \frac{12}{20} = \frac{17}{20}$
   $\frac{17}{20} + \frac{2}{3} = \frac{51}{60} + \frac{40}{60} = \frac{91}{60}$ or $1\frac{31}{60}$
7. $\frac{40}{48} > \frac{24}{48}$
8. $\frac{27}{45} > \frac{20}{45}$
9. $\frac{24}{32} = \frac{24}{32}$
10. $(600)\text{x}(60) = (36{,}000)$
    $558\text{x}62 = 34{,}596$
11. $(400)\text{x}(80) = (32{,}000)$
    $407\text{x}83 = 33{,}781$
12. $(300)\text{x}(10) = (3{,}000)$
    $349\text{x}12 = 4{,}188$
13. $128 \div 7 = 18\frac{2}{7}$
14. $471 \div 3 = 157$
15. $298 \div 5 = 59\frac{3}{5}$
16. $\frac{3}{1} \text{x} \frac{1}{4} = \frac{3}{4}$ ft
17. $\frac{4}{5} - \frac{1}{3} = \frac{12}{15} - \frac{5}{15} = \frac{7}{15}$ mile
18. $\frac{1}{3} \text{x} \frac{1}{2} = \frac{1}{6};\ \frac{1}{6}$ of 12 = 2 people
19. $13 + 19 + 26 = 58$ in
20. $30 \div 2 = 15$
    $15\text{x}1 = 15$ days

Lesson Practice 10A

1. done
2. done
3. $\frac{15}{24} \div \frac{16}{24} = \frac{15 \div 16}{1} = \frac{15}{16}$
4. $\frac{12}{16} \div \frac{8}{16} = \frac{12 \div 8}{1} = \frac{12}{8}$ or $1\frac{4}{8}$
5. $\frac{4}{8} \div \frac{2}{8} = \frac{4 \div 2}{1} = 2$
6. $\frac{1}{2} \div \frac{1}{8} = \frac{8}{16} \div \frac{2}{16} = \frac{8 \div 2}{1} = 4$ times
7. $\frac{2}{3} \div \frac{1}{6} = \frac{12}{18} \div \frac{3}{18} = \frac{12 \div 3}{1} = 4$ pieces
8. $\frac{6}{8} \div \frac{1}{4} = \frac{24}{32} \div \frac{8}{32} = \frac{24 \div 8}{1} = 3$ people

Lesson Practice 10B

1. $\frac{12}{20} \div \frac{5}{20} = \frac{12 \div 5}{1} = \frac{12}{5}$ or $2\frac{2}{5}$
2. $\frac{18}{24} \div \frac{4}{24} = \frac{18 \div 4}{1} = \frac{18}{4}$ or $4\frac{2}{4}$
3. $\frac{3}{12} \div \frac{4}{12} = \frac{3 \div 4}{1} = \frac{3}{4}$
4. $\frac{16}{24} \div \frac{15}{24} = \frac{16 \div 15}{1} = \frac{16}{15}$ or $1\frac{1}{15}$
5. $\frac{5}{20} \div \frac{16}{20} = \frac{5 \div 16}{1} = \frac{5}{16}$

6. $\frac{3}{4} \div \frac{1}{8} = \frac{24}{32} \div \frac{4}{32} = \frac{24 \div 4}{1} = 6$ times
7. $\frac{9}{10} \div \frac{1}{10} = \frac{9 \div 1}{1} = 9$ volunteers
8. $\frac{5}{16} \div \frac{1}{16} = \frac{5 \div 1}{1} = 5$ times

Lesson Practice 10C

1. $\frac{6}{15} \div \frac{10}{15} = \frac{6 \div 10}{1} = \frac{6}{10}$
2. $\frac{3}{6} \div \frac{2}{6} = \frac{3 \div 2}{1} = \frac{3}{2}$ or $1\frac{1}{2}$
3. $\frac{10}{15} \div \frac{12}{15} = \frac{10 \div 12}{1} = \frac{10}{12}$
4. $\frac{8}{16} \div \frac{8}{16} = \frac{8 \div 8}{1} = 1$
5. $\frac{5}{15} \div \frac{6}{15} = \frac{5 \div 6}{1} = \frac{5}{6}$
6. $\frac{1}{3} \div \frac{1}{9} = \frac{9}{27} \div \frac{3}{27} = \frac{9 \div 3}{1} = 3$ people
7. $\frac{4}{5} \div \frac{1}{10} = \frac{40}{50} \div \frac{5}{50} = \frac{40 \div 5}{1} = 8$ people
8. $\frac{1}{3} \div \frac{1}{6} = \frac{6}{18} \div \frac{3}{18} = \frac{6 \div 3}{1} = 2$ boards

Don't forget that the last step in these problems is optional for now.

Systematic Review10D

1. $\frac{10}{12} \div \frac{6}{12} = \frac{10 \div 6}{1} = \frac{10}{6}$ or $1\frac{4}{6}$
2. $\frac{4}{5} \div \frac{2}{5} = \frac{4 \div 2}{1} = 2$
3. $\frac{12}{15} \div \frac{5}{15} = \frac{12 \div 5}{1} = \frac{12}{5}$ or $2\frac{2}{5}$
4. $\frac{3}{4} \text{x} \frac{1}{8} = \frac{3}{32}$
5. $\frac{2}{3} \text{x} \frac{1}{6} = \frac{2}{18}$
6. $\frac{6}{8} \text{x} \frac{1}{4} = \frac{6}{32}$
7. $\frac{12}{18} - \frac{3}{18} = \frac{9}{18}$
8. $\frac{5}{10} - \frac{4}{10} = \frac{1}{10}$
9. $\frac{3}{6} + \frac{4}{6} = \frac{7}{6}$
   $\frac{7}{6} + \frac{4}{5} = \frac{35}{30} + \frac{24}{30} = \frac{59}{30}$ or $1\frac{29}{30}$
10. 38x94 = 3,572
11. 237x15 = 3,555
12. 709x51 = 36,159
13. done
14. (500) ÷ (50) = (10)
15. (600) ÷ (30) = (20)
16. $\frac{1}{2} \text{x} \frac{5}{8} = \frac{5}{16}$ of a pizza
17. $\frac{4}{6} \div \frac{1}{3} = \frac{12}{18} \div \frac{6}{18} = \frac{12 \div 6}{1} = 2$ pieces
18. $\frac{1}{2} + \frac{1}{6} = \frac{6}{12} + \frac{2}{12} = \frac{8}{12}$
    of the congregation

Systematic Review10E

1. $\frac{2}{3} \div \frac{1}{3} = \frac{2 \div 1}{1} = 2$
2. $\frac{15}{20} \div \frac{8}{20} = \frac{15 \div 8}{1} = \frac{15}{8}$ or $1\frac{7}{8}$
3. $\frac{8}{12} \div \frac{6}{12} = \frac{8 \div 6}{1} = \frac{8}{6}$ or $1\frac{2}{6}$
4. $\frac{4}{5} \text{x} \frac{2}{5} = \frac{8}{25}$
5. $\frac{6}{8} \text{x} \frac{1}{2} = \frac{6}{16}$
6. $\frac{1}{3} \text{x} \frac{2}{5} = \frac{2}{15}$
7. $\frac{9}{12} - \frac{8}{12} = \frac{1}{12}$
8. $\frac{8}{80} + \frac{70}{80} = \frac{78}{80}$
9. $\frac{9}{18} + \frac{4}{18} = \frac{13}{18}$
   $\frac{13}{18} + \frac{3}{4} = \frac{52}{72} + \frac{54}{72} = \frac{106}{72}$ or $1\frac{34}{72}$
10. 73 × 28 = 2,044
11. 829 × 72 = 59,688
12. 164 × 53 = 8,692
13. done
14. (700) ÷ (70) = (10)
15. (900) ÷ (20) ≈ (40 or 45)
16. 50 ÷ 2 = 25; 25x1=$25

17. $\frac{6}{42}+\frac{7}{42}=\frac{13}{42}$ of his trees

18. $\frac{6}{15}-\frac{5}{15}=\frac{1}{15}$ more

19. $\frac{24}{32}\div\frac{4}{32}=\frac{24\div4}{1}=6$ people

20. $\frac{1}{2}\times\frac{3}{4}=\frac{3}{8}$ cup

Systematic Review10F

1. $\frac{5}{8}\div\frac{6}{8}=\frac{5\div6}{1}=\frac{5}{6}$

2. $\frac{8}{12}\div\frac{3}{12}=\frac{8\div3}{1}=\frac{8}{3}$ or $2\frac{2}{3}$

3. $\frac{12}{24}\div\frac{4}{24}=\frac{12\div4}{1}=3$

4. $\frac{4}{5}\text{x}\frac{1}{10}=\frac{4}{50}$

5. $\frac{5}{6}\text{x}\frac{1}{12}=\frac{5}{72}$

6. $\frac{1}{2}\text{x}\frac{1}{4}=\frac{1}{8}$

7. $\frac{5}{20}+\frac{8}{20}=\frac{13}{20}$

8. $\frac{40}{50}-\frac{35}{50}=\frac{5}{50}$

9. $\frac{12}{22}+\frac{11}{22}=\frac{23}{22}$

   $\frac{23}{22}+\frac{2}{3}=\frac{69}{66}+\frac{44}{66}=\frac{113}{66}$ or $1\frac{47}{66}$

10. 35x16 = 560

11. 182x68 = 12,376

12. 390x41 = 15,990

13. (600) ÷ (40) = (15)

14. (400) ÷ (80) = (5)

15. (400) ÷ (10) = (40)

16. $\frac{1}{5}\text{x}\frac{2}{3}=\frac{2}{15}$ of the job

17. $\frac{1}{4}\div\frac{1}{16}=\frac{16}{64}\div\frac{4}{64}=\frac{16\div4}{1}=4$ people

18. $\frac{3}{7}+\frac{1}{4}=\frac{12}{28}+\frac{7}{28}=\frac{19}{28}$ of her time

19. 256x35 = 8,960 insects

20. $\frac{5}{5}-\frac{1}{5}=\frac{4}{5}$ only words

    $\frac{5}{6}\text{x}\frac{4}{5}=\frac{20}{30}$ of the pages read

Systematic Review11A

1. done
2. 1x10
   2x5
   1,2,5,10
3. 1x18
   2x9
   3x6
   1,2,3,6,9,18
4. yes
5. no
6. no
7. no
8. yes
9. yes
10. yes
11. yes
12. no
13. yes
14. no
15. yes
16. yes
17. yes
18. no
19. done
20. 1, 3, 5, 15
    1, 2, 5, 10
    GCF = 5
21. 1, 2, 3, 6
    1, 2, 3, 6, 9, 18
    GCF = 6

Lesson Practice 11B

1. 1x9
   3x3
   1,3,9
2. 1x4
   2x2
   1,2,4
3. 1x14
   2x7
   1,2,7,14
4. yes
5. yes
6. no
7. yes
8. no
9. no
10. done
11. 1x28
    2x14
    4x7
    1,2,4,7,14,28
12. 1, 3, 9
    1, 3, 5, 15
    GCF = 3
13. 1, 2, 5, 10
    1, 5, 25
    GCF = 5
14. 1, 2, 3, 6
    1, 2, 3, 4, 6, 12
    GCF = 6

Lesson Practice 11C

1. 1x8
   2x4
   1,2,4,8
2. 1x20
   2x10
   4x5
   1,2,4,5,10,20
3. 1x22
   2x11
   1,2,11,22
4. no
5. no
6. yes
7. yes
8. no
9. no
10. 1x42
    2x21
    3x14
    6x7
    1,2,3,6,7,14,21,42
11. 1x34
    2x17
    1,2,17,34
12. 1x50
    2x25
    5x10
    1,2,5,10,25,50
13. 1, 2, 4, 8
    1, 2, 3, 4, 6, 8, 12, 16, 24, 48
    GCF = 8
14. 1, 3, 5, 15
    1, 5, 7, 35
    GCF = 5
15. 1, 3, 9
    1, 2, 3, 6, 9, 18
    GCF = 9

Systematic Review 11D

1. yes
2. yes
3. no
4. 1, 2, 4, 7, 14, 28
   1, 2, 3, 6, 7, 14, 21, 42
   GCF = 14
5. 1, 2, 3, 6, 9, 18
   1, 3, 9, 27, 81
   GCF = 9

6. $\frac{1}{2} \times \frac{1}{3} = \frac{1}{6}$
7. $\frac{2}{3} \times \frac{4}{5} = \frac{8}{15}$
8. $\frac{3}{6} \times \frac{2}{5} = \frac{6}{30}$
9. $\frac{28}{35} \div \frac{5}{35} = \frac{28 \div 5}{1} = \frac{28}{5}$ or $5\frac{3}{5}$
10. $\frac{15}{24} \div \frac{8}{24} = \frac{15 \div 8}{1} = \frac{15}{8}$ or $1\frac{7}{8}$
11. $\frac{24}{30} \div \frac{5}{30} = \frac{24 \div 5}{1} = \frac{24}{5}$ or $4\frac{4}{5}$
12. done
13. $498 \div 51 = 9\frac{39}{51}$
14. $560 \div 28 = 20$
15. $\frac{1}{2} + \frac{1}{4} = \frac{4}{8} + \frac{2}{8} = \frac{6}{8}$ eaten

    $\frac{8}{8} - \frac{6}{8} = \frac{2}{8}$ left
16. $\frac{6}{30} + \frac{10}{30} = \frac{16}{30}$ of the class
17. $36 \div 3 = 12$;

    $3 \times 1 = 12$ girls

    $36 - 12 = 24$ boys
18. $\frac{4}{5} \times \frac{1}{2} = \frac{4}{10}$ of the total pay

Systematic Review 11E

1. yes
2. no
3. $\underline{1}, \underline{5}, 25$

    $\underline{1}, 2, 3, \underline{5}, 6, 10, 15, 30$

    GCF = 5
4. $\underline{1}, \underline{3}, 5, \underline{9}, 15, 45$

    $\underline{1}, \underline{3}, \underline{9}, 27$

    GCF = 9
5. $\frac{2}{3} \times \frac{7}{8} = \frac{14}{24}$
6. $\frac{1}{2} \times \frac{3}{9} = \frac{3}{18}$
7. $\frac{3}{4} \times \frac{5}{6} = \frac{15}{24}$
8. $\frac{20}{32} \div \frac{8}{32} = \frac{20 \div 8}{1} = \frac{20}{8}$ or $2\frac{4}{8}$
9. $\frac{36}{54} \div \frac{9}{54} = \frac{36 \div 9}{1} = 4$
10. $\frac{4}{8} \div \frac{6}{8} = \frac{4 \div 6}{1} = \frac{4}{6}$
11. $\frac{24}{56} < \frac{28}{56}$
12. $\frac{6}{15} > \frac{5}{15}$
13. $\frac{20}{36} < \frac{27}{36}$
14. $424 \div 53 = 8$
15. $711 \div 74 = 9\frac{45}{74}$
16. $890 \div 22 = 40\frac{10}{22}$
17. $25 \times 52 = 1{,}300$ people
18. $\frac{2}{8} + \frac{1}{8} + \frac{4}{8} = \frac{7}{8}$ completed

    $\frac{8}{8} - \frac{7}{8} = \frac{1}{8}$ to be finished
19. $\frac{24}{32} \div \frac{4}{32} = \frac{24 \div 4}{1} = 6$ sections
20. $\frac{1}{12} \times \frac{1}{4} = \frac{1}{48}$ of this year

Systematic Review 11F

1. no
2. no
3. $\underline{1}, \underline{2}, 4, 8, 16$

    $\underline{1}, \underline{2}, 17, 34$

    GCF = 2
4. $\underline{1}, \underline{2}, 3, \underline{4}, 6, 12$

    $\underline{1}, \underline{2}, \underline{4}, 5, 8, 10, 20, 40$

    GCF = 4
5. $\frac{5}{9} \times \frac{1}{2} = \frac{5}{18}$
6. $\frac{3}{5} \times \frac{1}{4} = \frac{3}{20}$
7. $\frac{6}{7} \times \frac{2}{3} = \frac{12}{21}$
8. $\frac{3}{27} \div \frac{9}{27} = \frac{3 \div 9}{1} = \frac{3}{9}$
9. $\frac{15}{18} \div \frac{12}{18} = \frac{15 \div 12}{1} = \frac{15}{12}$ or $1\frac{3}{12}$

10. $\frac{16}{32} \div \frac{8}{32} = \frac{16 \div 8}{1} = 2$
11. $\frac{6}{12} < \frac{10}{12}$
12. $\frac{40}{80} = \frac{40}{80}$
13. $\frac{9}{108} < \frac{24}{108}$
14. $559 \div 43 = 13$
15. $414 \div 79 = 5\frac{19}{79}$
16. $392 \div 14 = 28$
17. $14 + 14 + 14 + 14 = 56'$ perimeter
    $\frac{1}{7}$ of $56 = 8'$ for door openings
    $56' - 8' = 48'$ baseboard
18. yes
19. $\frac{4}{32} + \frac{8}{32} = \frac{12}{32}$ of his money
20. $\frac{1}{8}$ of $48 = 6$
    $\frac{1}{4}$ of $48 = 12$
    $6 + 12 = 18$
    $48 - 18 = \$30$

Lesson Practice 12A

1. done
2. $\frac{2}{4} \div \frac{2}{2} = \frac{1}{2}$
3. $\frac{18}{24} \div \frac{6}{6} = \frac{3}{4}$
4. $\frac{8}{12} \div \frac{4}{4} = \frac{2}{3}$
5. done
6. $\frac{6}{8} \div \frac{2}{2} = \frac{3}{4}$
7. $\frac{4}{8} \div \frac{4}{4} = \frac{1}{2}$
8. done
9. $\underline{1}, \underline{2}, \underline{4}$
   $\underline{1}, \underline{2}, 3, \underline{4}, 6, 8, 12, 24$
   GCF = 4
   $\frac{4}{24} \div \frac{4}{4} = \frac{1}{6}$
10. $\underline{1}, \underline{2}, \underline{3}, \underline{6}$
    $\underline{1}, \underline{2}, \underline{3}, \underline{6}, 9, 18$
    GCF = 6
    $\frac{6}{18} \div \frac{6}{6} = \frac{1}{3}$
11. $\underline{1}, \underline{2}, \underline{3}, \underline{6}, 9, 18$
    $\underline{1}, \underline{2}, \underline{3}, 5, \underline{6}, 10, 15, 30$
    GCF = 6
    $\frac{18}{30} \div \frac{6}{6} = \frac{3}{5}$

Lesson Practice 12B

1. $\frac{9}{12} \div \frac{3}{3} = \frac{3}{4}$
2. $\frac{4}{8} \div \frac{4}{4} = \frac{1}{2}$
3. $\frac{8}{10} \div \frac{2}{2} = \frac{4}{5}$
4. $\frac{25}{30} \div \frac{5}{5} = \frac{5}{6}$
5. $\frac{5}{20} \div \frac{5}{5} = \frac{1}{4}$
6. $\frac{10}{12} \div \frac{2}{2} = \frac{5}{6}$
7. $\frac{12}{18} \div \frac{6}{6} = \frac{2}{3}$
8. $\underline{1}, \underline{3}$
   $\underline{1}, \underline{3}, 5, 15$
   GCF = 3
   $\frac{3}{15} \div \frac{3}{3} = \frac{1}{5}$
9. $\underline{1}, \underline{2}, \underline{4}, 8$
   $\underline{1}, \underline{2}, 3, \underline{4}, 6, 12$
   GCF = 4
   $\frac{8}{12} \div \frac{4}{4} = \frac{2}{3}$
10. $\underline{1}, \underline{3}, 5, 15$
    $\underline{1}, 2, \underline{3}, 6, 9, 18$
    GCF = 3
    $\frac{15}{18} \div \frac{3}{3} = \frac{5}{6}$

11. $\underline{1}, 2, \underline{7}, 14$
    $\underline{1}, 3, \underline{7}, 21$
    GCF = 7
    $\frac{14}{21} \div \frac{7}{7} = \frac{2}{3}$

Lesson Practice 12C

1. $\frac{10}{15} \div \frac{5}{5} = \frac{2}{3}$
2. $\frac{9}{15} \div \frac{3}{3} = \frac{3}{5}$
3. $\frac{27}{33} \div \frac{3}{3} = \frac{9}{11}$
4. $\frac{20}{36} \div \frac{4}{4} = \frac{5}{9}$
5. $\frac{42}{48} \div \frac{6}{6} = \frac{7}{8}$
6. $\underline{1}, 2, \underline{5}, 10$
   $\underline{1}, 3, \underline{5}, 15$
   GCF = 5
   $\frac{10}{15} \div \frac{5}{5} = \frac{2}{3}$
7. $\underline{1}, \underline{2}, \underline{4}, \underline{8}, 16$
   $\underline{1}, \underline{2}, 3, \underline{4}, 6, \underline{8}, 12, 24$
   GCF = 8
   $\frac{16}{24} \div \frac{8}{8} = \frac{2}{3}$
8. $\frac{25}{30} \div \frac{5}{5} = \frac{5}{6}$
9. $\frac{6}{8} \div \frac{2}{2} = \frac{3}{4}$
10. $\frac{10}{16} \div \frac{2}{2} = \frac{5}{8}$
11. $\frac{18}{21} \div \frac{3}{3} = \frac{6}{7}$
12. $\frac{45}{50} \div \frac{5}{5} = \frac{9}{10}$
13. $\frac{27}{36} \div \frac{9}{9} = \frac{3}{4}$
14. $\frac{7}{14} \div \frac{7}{7} = \frac{1}{2}$
    $\frac{1}{2}$ of 56=28 people

Systematic Review 12D

1. $\frac{12}{28} \div \frac{4}{4} = \frac{3}{7}$
2. $\frac{14}{49} \div \frac{7}{7} = \frac{2}{7}$
3. $\frac{35}{50} \div \frac{5}{5} = \frac{7}{10}$
4. yes
5. yes
6. $\frac{24}{32} + \frac{4}{32} = \frac{28}{32} = \frac{7}{8}$
7. $\frac{5}{35} + \frac{14}{35} = \frac{19}{35}$
8. $\frac{6}{48} \div \frac{6}{6} = \frac{1}{8}$
9. $\frac{4}{50} \div \frac{2}{2} = \frac{2}{25}$
10. $\frac{8}{32} \div \frac{20}{32} = \frac{8 \div 20}{1} = \frac{8}{20} = \frac{2}{5}$
11. $\frac{6}{15} \div \frac{10}{15} = \frac{6 \div 10}{1} = \frac{6}{10} = \frac{3}{5}$
12. 531x624 = 331,344
13. 3,728x128 = 477,184
14. 6,593x756 = 4,984,308
15. $\frac{2}{4} \div \frac{2}{2} = \frac{1}{2}$ of an orange
16. $\frac{4}{6} \div \frac{1}{6} = \frac{4 \div 1}{1} = 4$ helpers
17. $\frac{3}{5} \times \frac{2}{3} = \frac{6}{15} = \frac{2}{5}$ of the letters
18. $\frac{2}{3} + \frac{1}{5} = \frac{10}{15} + \frac{3}{15} = \frac{13}{15}$ are correct

Systematic Review 12E

1. $\frac{24}{30} \div \frac{6}{6} = \frac{4}{5}$
2. $\frac{18}{28} \div \frac{2}{2} = \frac{9}{14}$
3. $\frac{15}{35} \div \frac{5}{5} = \frac{3}{7}$
4. no
5. no
6. $\frac{8}{16} + \frac{6}{16} = \frac{14}{16} = \frac{7}{8}$
7. $\frac{20}{50} + \frac{15}{50} = \frac{35}{50} = \frac{7}{10}$

8. $\frac{10}{24} = \frac{5}{12}$
9. $\frac{18}{32} = \frac{9}{16}$
10. $\frac{9}{54} \div \frac{30}{54} = \frac{9 \div 30}{1} = \frac{9}{30} = \frac{3}{10}$
11. $\frac{4}{8} \div \frac{6}{8} = \frac{4 \div 6}{1} = \frac{4}{6} = \frac{2}{3}$
12. $728 \text{x} 165 = 120{,}120$
13. $2{,}192 \text{x} 864 = 1{,}893{,}888$
14. $8{,}651 \text{x} 549 = 4{,}749{,}399$
15. $\frac{5}{15} \div \frac{5}{5} = \frac{1}{3}$
16. $512 \text{x} 150 = 76{,}800$ paper clips
17. $\frac{2}{3} + \frac{1}{5} = \frac{10}{15} + \frac{3}{15} = \frac{13}{15}$ of the beads
18. $360 \div 30 = 12$ paychecks
19. $\frac{3}{4} \text{x} \frac{1}{2} = \frac{3}{8}$
20. $\frac{3}{4} - \frac{3}{8} = \frac{24}{32} - \frac{12}{32} = \frac{12}{32} = \frac{3}{8}$ yd

Systematic Review 12F

1. $\frac{18}{63} \div \frac{9}{9} = \frac{2}{7}$
2. $\frac{26}{32} \div \frac{2}{2} = \frac{13}{16}$
3. $\frac{42}{48} \div \frac{6}{6} = \frac{7}{8}$
4. yes
5. yes
6. $\frac{36}{72} + \frac{8}{72} = \frac{44}{72} = \frac{11}{18}$
7. $\frac{20}{30} + \frac{6}{30} = \frac{26}{30} = \frac{13}{15}$
8. $\frac{12}{42} \div \frac{6}{6} = \frac{2}{7}$
9. $\frac{4}{20} \div \frac{4}{4} = \frac{1}{5}$
10. $\frac{8}{32} \div \frac{28}{32} = \frac{8 \div 28}{1} = \frac{8}{28} = \frac{2}{7}$
11. $\frac{3}{6} \div \frac{4}{6} = \frac{3 \div 4}{1} = \frac{3}{4}$
12. $371 \text{x} 244 = 90{,}524$
13. $5{,}970 \text{x} 186 = 1{,}110{,}420$
14. $7{,}035 \text{x} 369 = 2{,}595{,}915$
15. $\frac{18}{30} \div \frac{6}{6} = \frac{3}{5}$
    $500 \div 5 = 100$; $100 \text{x} 3 = 300$ people
16. $\frac{16}{60} > \frac{15}{60}$ so more sopranos
17. $16 \text{x} 12 = 192$; $192 \div 3 = 64$ packs
18. $\frac{3}{4}$ of 64=\$48
19. $\frac{12}{18} \div \frac{6}{18} = \frac{12 \div 6}{1} = 2$ pieces
20. $\frac{1}{3} \div \frac{4}{1} = \frac{1}{3} \div \frac{12}{3} = \frac{1}{12}$ of a pie

Lesson Practice 13A

1. done
2. $3 \times 3 \times 5$

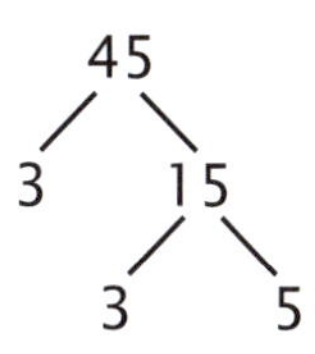

3. $2 \times 3 \times 5$

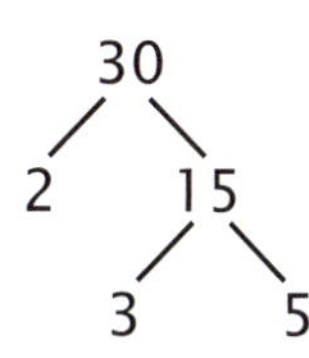

4. done
5. 3 x 3 x 3
6. 2 x 5 x 5
7. done
8. $\frac{20}{28} = \frac{2\text{x}2\text{x}5}{2\text{x}2\text{x}7} = \frac{2}{2} \text{x} \frac{2}{2} \text{x} \frac{5}{7} = \frac{5}{7}$
9. $\frac{30}{50} = \frac{2\text{x}3\text{x}5}{2\text{x}5\text{x}5} = \frac{2}{2} \text{x} \frac{5}{5} \text{x} \frac{3}{5} = \frac{3}{5}$
10. $\frac{21}{33} = \frac{3\text{x}7}{3\text{x}11} = \frac{3}{3} \text{x} \frac{7}{11} = \frac{7}{11}$
11. $\frac{18}{26} = \frac{2\text{x}3\text{x}3}{2\text{x}13} = \frac{2}{2} \text{x} \frac{3\text{x}3}{13} = \frac{9}{13}$
12. $\frac{24}{54} = \frac{2\text{x}2\text{x}2\text{x}3}{2\text{x}3\text{x}3\text{x}3} = \frac{2}{2} \text{x} \frac{3}{3} \text{x}$
    $\frac{2\text{x}2}{3\text{x}3} = \frac{4}{9}$

Lesson Practice 13B

1. $2 \times 5 \times 5 \times 5$

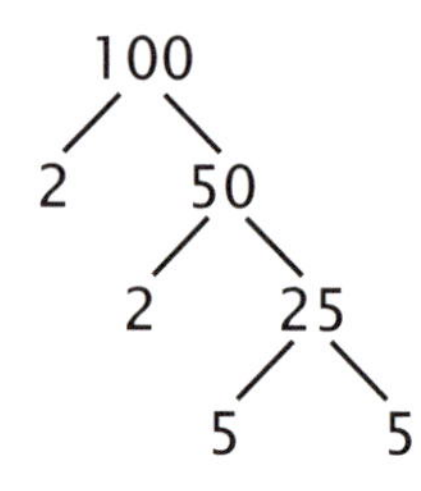

2. $2 \times 2 \times 2 \times 2 \times 2$

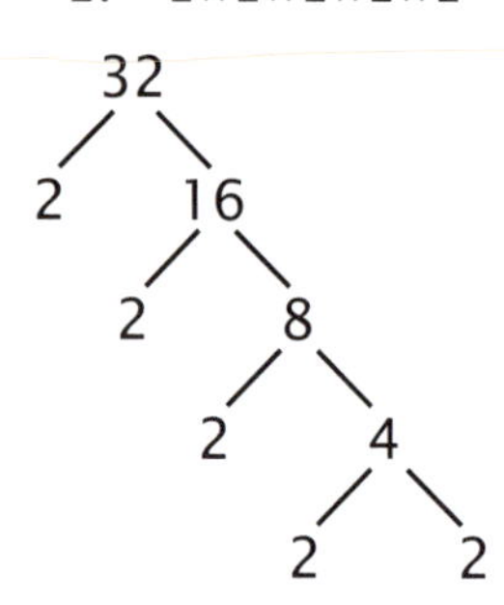

3. $2 \times 2 \times 2 \times 3 \times 3$

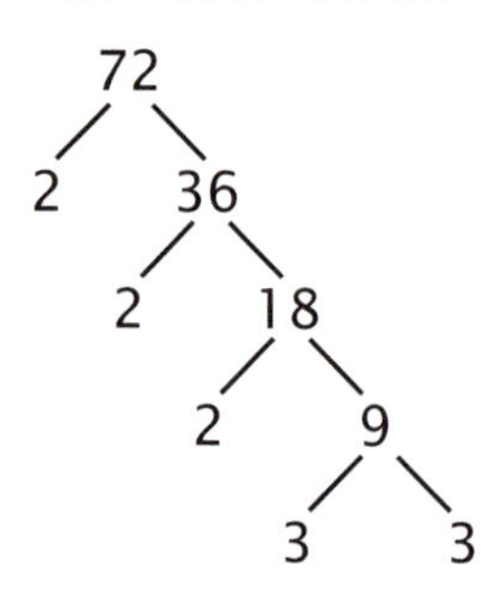

4. 2 x 2 x 2 x 2 x 3
5. 2 x 2 x 7
6. 2 x 3 x 7
7. $\frac{45}{75} = \frac{3x3x5}{3x5x5} = \frac{3}{3} x \frac{5}{5} x \frac{3}{5} = \frac{3}{5}$
8. $\frac{42}{60} = \frac{2x3x7}{2x2x3x5} = \frac{2}{2} x \frac{3}{3} x \frac{7}{2x5} = \frac{7}{10}$
9. done
10. $\frac{33}{44} = \frac{3x\cancel{11}}{2x2x\cancel{11}} = \frac{3}{4}$
11. $\frac{40}{90} = \frac{\cancel{2}x2x2x\cancel{5}}{\cancel{2}x3x3x\cancel{5}} = \frac{4}{9}$
12. $\frac{27}{48} = \frac{\cancel{3}x3x3}{2x2x2x2x\cancel{3}} = \frac{9}{16}$

Lesson Practice 13C

1. $2 \times 2 \times 5$

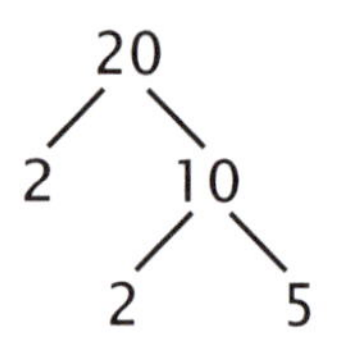

2. $3 \times 3 \times 3 \times 3$

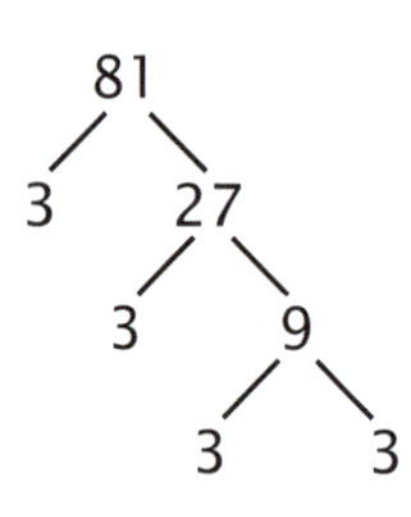

3. $2 \times 2 \times 13$

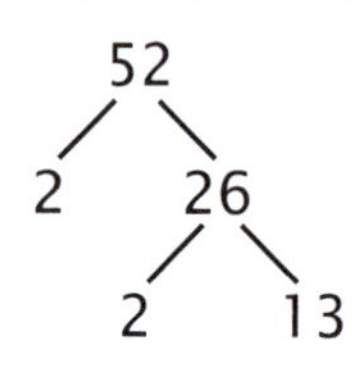

4. 2 x 3 x 11
5. 2 x 2 x 11
6. 2 x 2 x 2 x 2 x 5
7. $\frac{20}{24} = \frac{\cancel{2}x\cancel{2}x5}{\cancel{2}x\cancel{2}x2x3} = \frac{5}{6}$
8. $\frac{30}{36} = \frac{\cancel{2}x\cancel{3}x5}{\cancel{2}x2x\cancel{3}x3} = \frac{5}{6}$
9. $\frac{48}{54} = \frac{\cancel{2}x2x2x2x\cancel{3}}{\cancel{2}x\cancel{3}x3x3} = \frac{8}{9}$
10. $\frac{9}{27} = \frac{\cancel{3}x\cancel{3}}{3x\cancel{3}x\cancel{3}} = \frac{1}{3}$
11. $\frac{15}{25} = \frac{3x\cancel{5}}{5x\cancel{5}} = \frac{3}{5}$
12. $\frac{12}{15} = \frac{2x2x\cancel{3}}{\cancel{3}x5} = \frac{4}{5}$

Systematic Review 13D

1. 2x13
2. 2x2x3x5

3. $2\times2\times2\times3$
4. $\frac{18}{24}=\frac{\cancel{2}\times\cancel{3}\times3}{\cancel{2}\times2\times2\times\cancel{3}}=\frac{3}{4}$
5. $\frac{15}{30}=\frac{\cancel{3}\times\cancel{5}}{2\times\cancel{3}\times\cancel{5}}=\frac{1}{2}$
6. $\frac{8}{32}+\frac{8}{32}=\frac{16}{32}\div\frac{16}{16}=\frac{1}{2}$
7. $\frac{10}{16}-\frac{8}{16}=\frac{2}{16}\div\frac{2}{2}=\frac{1}{8}$
8. $\frac{6}{21}+\frac{7}{21}=\frac{13}{21}$
9. $\frac{2}{12}\div\frac{6}{12}=\frac{2\div6}{1}=\frac{2}{6}\div\frac{2}{2}=\frac{1}{3}$
10. $\frac{5}{9}\times\frac{3}{7}=\frac{15}{63}\div\frac{3}{3}=\frac{5}{21}$
11. $\frac{8}{11}\times\frac{3}{4}=\frac{24}{44}\div\frac{4}{4}=\frac{6}{11}$
12. done
13. $9442\div53=178\frac{8}{53}$
14. $4925\div189=26\frac{11}{189}$
15. $\frac{1}{6}\times12=2$ months
16. $\frac{3}{4}+\frac{1}{8}=\frac{24}{32}+\frac{4}{32}=\frac{28}{32}\div\frac{4}{4}=\frac{7}{8}$ yd
17. $2{,}845\div5=569$ gal
18. $528\times73=38{,}544$ mosquitoes

Systematic Review 13E

1. $3\times3\times3\times3$
2. $2\times3\times3\times5$
3. $3\times5\times5$
4. $\frac{8}{22}=\frac{\cancel{2}\times2\times2}{\cancel{2}\times11}=\frac{4}{11}$
5. $\frac{32}{48}=\frac{\cancel{2}\times\cancel{2}\times\cancel{2}\times\cancel{2}\times2}{\cancel{2}\times\cancel{2}\times\cancel{2}\times\cancel{2}\times3}=\frac{2}{3}$
6. $\frac{6}{12}+\frac{4}{12}=\frac{10}{12}\div\frac{2}{2}=\frac{5}{6}$
7. $\frac{18}{27}-\frac{9}{27}=\frac{9}{27}\div\frac{9}{9}=\frac{1}{3}$
8. $\frac{30}{50}+\frac{5}{50}=\frac{35}{50}\div\frac{5}{5}=\frac{7}{10}$
9. $\frac{16}{64}\div\frac{48}{64}=\frac{16\div48}{1}=\frac{16}{48}\div\frac{16}{16}=\frac{1}{3}$
10. $\frac{5}{9}\times\frac{2}{6}=\frac{10}{54}\div\frac{2}{2}=\frac{5}{27}$
11. $\frac{2}{5}\times\frac{3}{4}=\frac{6}{20}\div\frac{2}{2}=\frac{3}{10}$
12. $\frac{30}{48}<\frac{32}{48}$
13. $\frac{7}{21}>\frac{6}{21}$
14. $\frac{40}{80}=\frac{40}{80}$
15. $1130\div38=29\frac{28}{38}$ or $29\frac{14}{19}$
16. $2686\div22=122\frac{2}{22}$ or $122\frac{1}{11}$
17. $5032\div235=21\frac{97}{235}$
18. $45\times365=16{,}425$ pennies
19. $\frac{12}{36}+\frac{15}{36}=\frac{27}{36}$ of an hour
20. $\frac{27}{36}$ of 60 : $\frac{27}{36}\times\frac{60}{1}=\frac{1{,}620}{36}=45$ min
    $\frac{3}{4}$ of 60 : $60\div4=15$; $15\times3=45$ min
    The reduced fraction is easier to use.

Systematic Review 13F

1. $2\times2\times2\times2\times2\times2$
2. $2\times2\times2\times2$
3. $3\times3\times5$
4. $\frac{81}{90}=\frac{\cancel{3}\times\cancel{3}\times3\times3}{2\times\cancel{3}\times\cancel{3}\times5}=\frac{9}{10}$
5. $\frac{12}{18}=\frac{\cancel{2}\times2\times\cancel{3}}{\cancel{2}\times\cancel{3}\times3}=\frac{2}{3}$
6. $\frac{18}{48}+\frac{8}{48}=\frac{26}{48}\div\frac{2}{2}=\frac{13}{24}$
7. $\frac{45}{50}-\frac{20}{50}=\frac{25}{50}\div\frac{25}{25}=\frac{1}{2}$
8. $\frac{12}{24}+\frac{6}{24}=\frac{18}{24}\div\frac{6}{6}=\frac{3}{4}$
9. $\frac{10}{50}\div\frac{20}{50}=\frac{10\div20}{1}=\frac{10}{20}\div\frac{10}{10}=\frac{1}{2}$
10. $\frac{4}{7}\times\frac{2}{8}=\frac{8}{56}\div\frac{8}{8}=\frac{1}{7}$
11. $\frac{5}{12}\times\frac{3}{5}=\frac{15}{60}\div\frac{15}{15}=\frac{1}{4}$

12. $\frac{60}{110} < \frac{77}{110}$
13. $\frac{9}{36} < \frac{16}{36}$
14. $\frac{15}{36} > \frac{12}{36}$
15. $3621 \div 96 = 37\frac{69}{96}$ or $37\frac{23}{32}$
16. $7192 \div 73 = 98\frac{38}{73}$
17. $6831 \div 120 = 56\frac{111}{120}$ or $56\frac{37}{40}$
18. $2{,}075 \div 25 = 83$ days
19. $\frac{6}{18} + \frac{3}{18} = \frac{9}{18} = \frac{1}{2}$
    $\frac{1}{2} + \frac{1}{8} = \frac{8}{16} + \frac{2}{16} = \frac{10}{16} = \frac{5}{8}$ of the job
20. $\frac{8}{8} - \frac{5}{8} = \frac{3}{8}$ of the job left

Lesson Practice 14A

1. done
2. $\frac{7}{10}$
3. $\frac{2}{3}$
4. $\frac{3}{7}$
5. done
6.

7. done
8. $\frac{12}{16} = \frac{3}{4}$
9. $\frac{8}{16} = \frac{1}{2}$
10. $\frac{3}{16}$
11. $\frac{14}{16} = \frac{7}{8}$
12. $\frac{10}{16} = \frac{5}{8}$

Lesson Practice 14B

1. $\frac{1}{3}$
2. $\frac{5}{8}$
3. $\frac{5}{6}$
4. $\frac{3}{10}$
5.

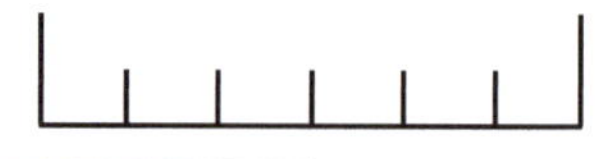

6.

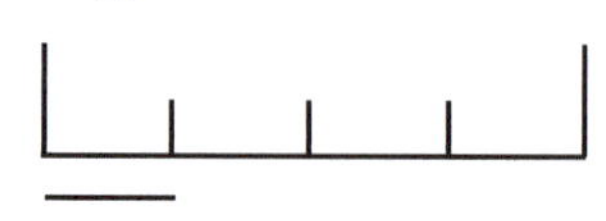

7. $\frac{1}{16}$
8. $\frac{4}{16} = \frac{1}{4}$
9. $\frac{7}{16}$
10. $\frac{16}{16} = 1$
11. $\frac{15}{16}$
12. $\frac{8}{16} = \frac{1}{2}$

Lesson Practice 14C

1. $\frac{6}{10}$
2. $\frac{4}{7}$
3. $\frac{3}{6}$
4. $\frac{1}{4}$
5.

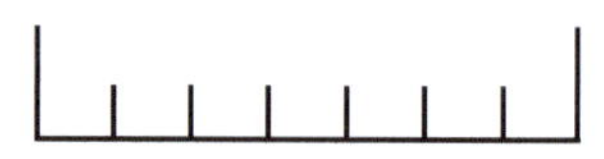

6.

7. $\frac{11}{16}$

8. $\frac{2}{16} = \frac{1}{8}$

9. $\frac{6}{16} = \frac{3}{8}$

10. $\frac{9}{16}$

11. $\frac{13}{16}$

12. $\frac{12}{16} = \frac{3}{4}$

Systematic Review 14D

1. $\frac{7}{8}$
2. $\frac{3}{6} = \frac{1}{2}$
3. $\frac{5}{16}$
4. $\frac{14}{16} = \frac{7}{8}$
5. 2x2x7
6. 5x11
7. 2x2x3x7
8. $\frac{48}{64} =$ $\frac{\cancel{2}x\cancel{2}x\cancel{2}x\cancel{2}x3}{\cancel{2}x\cancel{2}x\cancel{2}x\cancel{2}x2x2} = \frac{3}{4}$
9. done
10. 20x10 = 200 sq. ft
11. 15x7 = 105 sq. yd
12. no
13. yes
14. 12 : $\underline{2}, \underline{3}$, 4, $\underline{6}$, 12
    18 : $\underline{2}, \underline{3}, \underline{6}$, 9, 18
    GCF = 6
15. 25 : $\underline{5}$, $\underline{25}$
    50 : 2, $\underline{5}$, $\underline{25}$
    GCF = 25
16. 15x13 = 195 sq. ft

Systematic Review 14E

1. $\frac{2}{6} = \frac{1}{3}$
2. $\frac{6}{7}$
3. $\frac{10}{16} = \frac{5}{8}$
4. $\frac{3}{16}$
5. 2x5x7
6. 3x3x5
7. 2x3x5
8. $\frac{33}{63} = \frac{\cancel{3}x11}{\cancel{3}x3x7} = \frac{11}{21}$
9. 8x5 = 40 sq in
10. 90x45 = 4,050 sq. ft
11. 28x15 = 420 sq. yd
12. no
13. 21: $\underline{3}$, $\underline{7}$, $\underline{21}$
    42: 2, $\underline{3}$, 6, $\underline{7}$, 14, $\underline{21}$, 42
    GCF = 21
14. $\frac{3}{4} \times \frac{1}{3} = \frac{3}{12} \div \frac{3}{3} = \frac{1}{4}$ for school clothes
15. $\frac{8}{12} < \frac{9}{12}$ so $\frac{2}{3} < \frac{3}{4}$
16. $\frac{9}{12} - \frac{8}{12} = \frac{1}{12}$ of a cup
17. 15 + 13 + 15 + 13 = 56 ft
18. 56 ÷ 4 = 14
    14x3 = 42
    56 − 42 = 14 ft needed

Systematic Review 14F

1. $\frac{2}{3}$
2. $\frac{4}{10} = \frac{2}{5}$
3. $\frac{7}{16}$
4. $\frac{16}{16} = 1$

5. 2x3x11
6. 2x2x7
7. 2x3x3x3
8. $\frac{75}{100} = \frac{3 \times \cancel{5} \times \cancel{5}}{2 \times 2 \times \cancel{5} \times \cancel{5}} = \frac{3}{4}$
9. 3x2 = 6 sq. in
10. 11x6 = 66 sq. ft
11. 35x19 = 665 sq. yd
12. yes
13. $24: \underline{2}, \underline{3}, \underline{4}, \underline{6}, 8, \underline{12}, 24$

    $36: \underline{2}, \underline{3}, \underline{4}, \underline{6}, 9, \underline{12}, 18, 36$

    GCF = 12
14. $\frac{18}{42} + \frac{7}{42} = \frac{25}{42}$ of the chocolates; no
15. $\frac{8}{12} \div \frac{1}{6} = \frac{48}{72} \div \frac{12}{72} =$

    $\frac{48 \div 12}{1} = 4$ times
16. 36x12 = 432 eggs
17. 2,160 ÷ 432 = 5 crates
18. area

Lesson Practice 15A

1. done
2. $\frac{6}{6} + \frac{5}{6} = \frac{11}{6}$
3. $\frac{2}{2} + \frac{2}{2} + \frac{1}{2} = \frac{5}{2}$
4. done
5. $1 + 1 + \frac{1}{3} = 2\frac{1}{3}$
6. $1 + \frac{4}{5} = 1\frac{4}{5}$

Lesson Practice 15B

1. $\frac{4}{4} + \frac{1}{4} = \frac{5}{4}$
2. $\frac{6}{6} + \frac{6}{6} + \frac{2}{6} = \frac{14}{6}$
3. done
4. $\frac{3}{3} + \frac{3}{3} + \frac{3}{3} + \frac{2}{3} = \frac{11}{3}$
5. $1 + 1 + \frac{3}{5} = 2\frac{3}{5}$
6. done
7. $\frac{2}{2} + \frac{1}{2} = 1\frac{1}{2}$
8. $\frac{6}{6} + \frac{6}{6} + \frac{6}{6} + \frac{5}{6} =$

    $1 + 1 + 1 + \frac{5}{6} = 3\frac{5}{6}$

Lesson Practice 15C

1. $\frac{6}{6} + \frac{6}{6} + \frac{3}{6} = \frac{15}{6}$
2. $\frac{2}{2} + \frac{2}{2} + \frac{1}{2} = \frac{5}{2}$
3. $\frac{5}{5} + \frac{4}{5} = \frac{9}{5}$
4. $\frac{8}{8} + \frac{8}{8} + \frac{8}{8} + \frac{5}{8} = \frac{29}{8}$
5. $\frac{4}{4} + \frac{4}{4} + \frac{1}{4} = \frac{9}{4}$
6. $1 + 1 + \frac{2}{3} = 2\frac{2}{3}$
7. done
8. $\frac{7}{7} + \frac{1}{7} = 1\frac{1}{7}$
9. $\frac{2}{2} + \frac{2}{2} + \frac{1}{2} = 2\frac{1}{2}$
10. $\frac{4}{4} + \frac{4}{4} + \frac{4}{4} + \frac{4}{4} + \frac{1}{4} = 4\frac{1}{4}$

Systematic Review 15D

1. $\frac{3}{3} + \frac{3}{3} + \frac{2}{3} = \frac{8}{3}$
2. $\frac{2}{2} + \frac{2}{2} + \frac{2}{2} + \frac{1}{2} = \frac{7}{2}$
3. $\frac{9}{9} + \frac{4}{9} = 1\frac{4}{9}$
4. $\frac{5}{5} + \frac{5}{5} + \frac{1}{5} = 2\frac{1}{5}$
5. $\frac{15}{16}$
6. $\frac{4}{16} = \frac{1}{4}$

7. $\frac{16}{24} \div \frac{8}{8} = \frac{2}{3}$
8. $\frac{28}{35} \div \frac{7}{7} = \frac{4}{5}$
9. $\frac{54}{63} \div \frac{9}{9} = \frac{6}{7}$
10. done
11. done
12. $10 \times 10 = 100$ sq. in
13. $10+10+10+10 = 40$ in
14. $5 \times 5 = 25$ sq. mi
15. $5+5+5+5 = 20$ mi
16. 3x31
17. 44: 2, 4, $\underline{11}$, 22, 44
    55: 5, $\underline{11}$, 55
    GCF = 11
18. $\frac{2}{16} = \frac{1}{8}$
    $\frac{7}{8} \div \frac{1}{8} = \frac{7 \div 1}{1} = 7$ sections

Systematic Review 15E

1. $\frac{6}{6} + \frac{5}{6} = \frac{11}{6}$
2. $\frac{9}{9} + \frac{9}{9} + \frac{9}{9} + \frac{4}{9} = \frac{31}{9}$
3. $\frac{8}{8} + \frac{7}{8} = 1\frac{7}{8}$
4. $\frac{10}{10} + \frac{10}{10} + \frac{10}{10} + \frac{3}{10} = 3\frac{3}{10}$
5. $\frac{9}{16}$
6. $\frac{12}{16} = \frac{3}{4}$
7. $\frac{27}{45} \div \frac{9}{9} = \frac{3}{5}$
8. $\frac{44}{60} \div \frac{4}{4} = \frac{11}{15}$
9. $\frac{24}{30} \div \frac{6}{6} = \frac{4}{5}$
10. done
11. $11^2 = 11 \times 11 = 121$ sq. mi
12. $14^2 = 14 \times 14 = 196$ sq. in
13. 3x3x3x3
14. 18: $\underline{2}$, 3, 6, 9, 18
    28: $\underline{2}$, 4, 7, 14, 28
    GCF = 2
15. no
16. $\frac{1}{3} \times \frac{5}{6} = \frac{5}{18}$ of the people
17. $35x175 = $6,125
18. $12+12+12+12 = 48$ mi

Systematic Review 15F

1. $\frac{8}{8} + \frac{3}{8} = \frac{11}{8}$
2. $\frac{7}{7} + \frac{7}{7} + \frac{7}{7} + \frac{2}{7} = \frac{23}{7}$
3. $\frac{7}{7} + \frac{4}{7} = 1\frac{4}{7}$
4. $\frac{6}{6} + \frac{6}{6} + \frac{6}{6} + \frac{6}{6} + \frac{4}{6} = 4\frac{4}{6}$
5. $\frac{6}{16} = \frac{3}{8}$
6. $\frac{5}{16}$
7. $\frac{50}{70} \div \frac{10}{10} = \frac{5}{7}$
8. $\frac{24}{36} \div \frac{12}{12} = \frac{2}{3}$
9. $\frac{3}{9} \div \frac{3}{3} = \frac{1}{3}$
10. $25^2 = 625$ sq. ft
11. $2^2 = 4$ sq. in
12. $17^2 = 289$ sq. ft
13. done
14. $3^2 = 9$
15. $10^2 = 100$
16. 21: $\underline{3}$, 7, 21
    24: 2, $\underline{3}$, 4, 6, 8, 12, 24
    GCF = $\underline{3}$
17. yes
18. 2x5x5
19. $\frac{3}{6} + \frac{4}{6} = \frac{7}{6} = 1\frac{1}{6}$ pies
20. $8{,}925 \div 425 = 21$ days

Lesson Practice 16A

1. done
2. $4\frac{3}{10}$
3. $2\frac{1}{8}$
4. $1\frac{5}{8}$
5. $\frac{3}{8}$
6. $4\frac{3}{16}$

Lesson Practice 16B

1. $1\frac{4}{5}$
2. $3\frac{1}{5}$
3. $4\frac{3}{4}$
4. $2\frac{7}{16}$
5. $1\frac{1}{2}$
6. $3\frac{1}{4}$

Lesson Practice 16C

1. $2\frac{4}{10} = 2\frac{2}{5}$
2. $1\frac{6}{10} = 1\frac{3}{5}$
3. $\frac{1}{2}$
4. $3\frac{9}{16}$
5. $4\frac{3}{4}$
6. $2\frac{7}{8}$

Systematic Review 16D

1. $3\frac{9}{10}$
2. $\frac{8}{8}+\frac{8}{8}+\frac{6}{8}=\frac{22}{8}$
3. $\frac{6}{6}+\frac{6}{6}+\frac{6}{6}+\frac{1}{6}=\frac{19}{6}$
4. $\frac{5}{5}+\frac{2}{5}=1\frac{2}{5}$
5. $\frac{8}{8}+\frac{8}{8}+\frac{3}{8}=2\frac{3}{8}$
6. $\frac{15}{30}\div\frac{15}{15}=\frac{1}{2}$
7. $\frac{28}{70}\div\frac{14}{14}=\frac{2}{5}$
8. $\frac{6}{9}\div\frac{3}{3}=\frac{2}{3}$
9. $\frac{15}{40}+\frac{16}{40}=\frac{31}{40}$
10. $\frac{10}{14}-\frac{7}{14}=\frac{3}{14}$
11. $\frac{4}{8}+\frac{6}{8}+\frac{7}{8}=\frac{17}{8}=2\frac{1}{8}$
12. done
13. $(\frac{1}{2})(9)(12) = 54$ sq. in
14. $(\frac{1}{2})(30)(7) = 105$ sq. ft
15. 3x5x5
16. $15^2 = 225$ sq. ft
17. $\frac{1}{2} \text{x} \frac{4}{10} = \frac{4}{20} = \frac{1}{5}$ of the house
18. $3{,}279 \div 7 = 468\frac{3}{7}$ weeks

Systematic Review 16E

1. $4\frac{1}{4}$
2. $\frac{10}{10}+\frac{10}{10}+\frac{1}{10}=\frac{21}{10}$
3. $\frac{3}{3}+\frac{3}{3}+\frac{3}{3}+\frac{2}{3}=\frac{11}{3}$
4. $\frac{9}{9}+\frac{8}{9}=1\frac{8}{9}$
5. $\frac{2}{2}+\frac{2}{2}+\frac{1}{2}=2\frac{1}{2}$

6. $\frac{20}{32} \div \frac{8}{32} = \frac{20 \div 8}{1} = \frac{20}{8} = \frac{5}{2} = 2\frac{1}{2}$
7. $\frac{10}{14} \div \frac{7}{14} = \frac{10 \div 7}{1} = \frac{10}{7} = 1\frac{3}{7}$
8. $\frac{6}{8} \div \frac{4}{8} = \frac{6 \div 4}{1} = \frac{6}{4} = \frac{3}{2} = 1\frac{1}{2}$
9. $\frac{1}{5} = \frac{2}{10} = \frac{3}{15}$
10. $\frac{3}{4} = \frac{6}{8} = \frac{9}{12}$
11. $\frac{4}{7} = \frac{8}{14} = \frac{12}{21}$
12. $(\frac{1}{2})(6)(4) = 12$ sq. ft
13. $(\frac{1}{2})(8)(6) = 24$ sq. in
14. $\frac{1}{2}(21)(5) = 52\frac{1}{2}$ sq. ft
15. 15 : 3, $\underline{5}$,15
    20 : 2, 4, $\underline{5}$,10, 20
    GCF = 5
16. 15x15 = 225
17. $\frac{1}{2} + \frac{1}{2} + \frac{1}{2} = \frac{3}{2} = 1\frac{1}{2}$ in
18. 256x38 = 9,728 sq. yd
19. $\frac{3}{15} \div \frac{3}{3} = \frac{1}{5}$ of her income
20. $\frac{60}{72} < \frac{66}{72}$;
    Danny has mowed more of the lawn

Systematic Review 16F

1. $2\frac{3}{8}$
2. $\frac{4}{4} + \frac{4}{4} + \frac{3}{4} = \frac{11}{4}$
3. $\frac{7}{7} + \frac{7}{7} + \frac{7}{7} + \frac{5}{7} = \frac{26}{7}$
4. $\frac{6}{6} + \frac{5}{6} = 1\frac{5}{6}$
5. $\frac{10}{10} + \frac{10}{10} + \frac{7}{10} = 2\frac{7}{10}$
6. $\frac{1}{5} \times \frac{5}{6} = \frac{5}{30} \div \frac{5}{5} = \frac{1}{6}$
7. $\frac{2}{3} \times \frac{4}{5} = \frac{8}{15}$
8. $\frac{1}{3} \times \frac{2}{8} = \frac{2}{24} \div \frac{2}{2} = \frac{1}{12}$
9. $\frac{8}{12} < \frac{9}{12}$
10. $\frac{21}{56} > \frac{8}{56}$
11. $\frac{55}{99} > \frac{36}{99}$
12. 9x3 = 27 sq ft
13. 9+3+9+3 = 24 ft
14. $17^2 = 289$ sq. in
15. 17+17+17+17 = 68 in
16. $(\frac{1}{2})(16)(12) = 96$ sq. ft
17. 12+16+20 = 48 ft
18. 2x2x5x5
19. 20x20 = 400
20. $\frac{80}{128} \div \frac{8}{128} = \frac{80 \div 8}{1} = 10$ pieces

Lesson Practice 17A

1. done
2. $(3)+(3) = (6); 5\frac{5}{6}$
3. $(7)+(9) = (16); 16\frac{4}{5}$
4. $(2)+(2) = (4);\ 3\frac{8}{9}$
5. $(4)+(9) = (13);\ 12\frac{6}{7}$
6. $(7)+(7) = (14);\ 13\frac{3}{4}$
7. done
8. $(7)-(3) = (4);\ 3\frac{3}{8}$
9. $(11)-(10) = (1);\ 1\frac{1}{5}$
10. $(9)-(5) = (4);\ 4\frac{2}{9}$
11. $(6)-(4) = (2);\ 2\frac{1}{4}$
12. $(4)-(3) = (1);\ 1\frac{1}{6}$

Lesson Practice 17B

1. $(10)+(9)=(19);\ 18\frac{5}{6}$
2. $(8)+(4)=(12);\ 11\frac{7}{8}$
3. $(7)+(4)=(11);\ 11\frac{2}{3}$
4. $(6)+(4)=(10);\ 9\frac{3}{4}$
5. $(2)+(5)=(7);\ 6\frac{4}{5}$
6. $(10)+(4)=(14);\ 13\frac{6}{7}$
7. $(3)-(1)=(2);\ 1\frac{2}{5}$
8. $(9)-(6)=(3);\ 2\frac{3}{8}$
9. $(6)-(4)=(2);\ 1\frac{1}{5}$
10. $(5)-(3)=(2);\ 1\frac{4}{7}$
11. $(10)-(8)=(2);\ 2\frac{1}{6}$
12. $(17)-(8)=(9);\ 8\frac{4}{9}$
13. $2\frac{1}{3}+1\frac{1}{3}=3\frac{2}{3}$ pies
14. $4\frac{3}{5}-3\frac{1}{5}=1\frac{2}{5}$ lb

Lesson Practice 17C

1. $(5)+(8)=(13);\ 12\frac{7}{8}$
2. $(4)+(4)=(8);\ 7\frac{8}{9}$
3. $(8)+(6)=(14);\ 13\frac{3}{4}$
4. $(2)+(6)=(8);\ 8\frac{7}{10}$
5. $(9)+(1)=(10);\ 10\frac{3}{5}$
6. $(6)+(3)=(9);\ 9\frac{4}{9}$
7. $(13)-(4)=(9);\ 8\frac{2}{7}$
8. $(8)-(5)=(3);\ 3\frac{1}{8}$
9. $(15)-(9)=(6);\ 5\frac{2}{5}$
10. $(2)-(1)=(1);\ 1\frac{2}{11}$
11. $(8)-(3)=(5);\ 4\frac{1}{3}$
12. $(16)-(9)=(7);\ 6\frac{5}{8}$
13. $4\frac{7}{10}-1\frac{4}{10}=3\frac{3}{10}$ dollars
14. $1\frac{1}{8}+1\frac{2}{8}=2\frac{3}{8}$ mi

Systematic Review 17D

1. $3\frac{3}{7}+2\frac{1}{7}=5\frac{4}{7}$
2. $11\frac{3}{4}-8\frac{2}{4}=3\frac{1}{4}$
3. $7\frac{1}{5}+2\frac{3}{5}=9\frac{4}{5}$
4. $3\frac{4}{5}=\frac{19}{5}$
   (As you may have discovered, a shortcut is to think, "5x3=15 and 15+4=19, so $\frac{19}{15}$")
5. $4\frac{1}{8}=\frac{33}{8}$
6. $5\frac{5}{6}=\frac{35}{6}$
7. $\frac{19}{9}=2\frac{1}{9}$
   Remember that a fraction is also a division problem. Dividing and writing the answer with a fractional remainder produces a mixed number.
8. $\frac{24}{5}=4\frac{4}{5}$
9. $\frac{15}{8}=1\frac{7}{8}$
10. done
11. $20\text{x}20\text{x}20=8{,}000$ cu. ft
12. $12\text{x}12\text{x}12=1{,}728$ cu. ft
13. $10\text{x}10\text{x}10=1{,}000$ cu. ft

14. $7\frac{2}{3}-3\frac{1}{3}=4\frac{1}{3}$ mi
15. 2x3x5x5
16. $4^2=16$
17. 9x10 = 90 sq. ft
    90<100; yes
18. $\frac{1}{6}$x18 = 3 robins

Systematic Review 17E

1. $2\frac{4}{5}-1\frac{2}{5}=1\frac{2}{5}$
2. $4\frac{1}{9}+6\frac{7}{9}=10\frac{8}{9}$
3. $5\frac{5}{6}-2\frac{4}{6}=3\frac{1}{6}$
4. $2\frac{1}{2}=\frac{5}{2}$
5. $6\frac{2}{4}=\frac{26}{4}$
6. $5\frac{3}{8}=\frac{43}{8}$
7. $\frac{51}{10}=5\frac{1}{10}$
8. $\frac{55}{9}=6\frac{1}{9}$
9. $\frac{16}{5}=3\frac{1}{5}$
10. $1\frac{1}{4}$
11. 4x4x4 = 64 cu. in
12. 15x15x15 = 3,375 cu. ft
13. 30x30x30 = 27,000 cu. ft
14. 1x1x1 = 1 cu. ft
15. 25: $\underline{5}$, 25
    30: 2, 3, $\underline{5}$, 6, 10, 15, 30
    GCF = 5
16. $4\frac{1}{4}+3\frac{2}{4}=7\frac{3}{4}$ tons
17. $\frac{1}{3}+\frac{2}{3}=\frac{3}{3}=1$
    $1+\frac{3}{5}=1\frac{3}{5}$ yd
18. $\frac{4}{5}\text{x}\frac{1}{2}=\frac{4}{10}=\frac{2}{5}$ of the job

Systematic Review 17F

1. $4\frac{2}{5}+8\frac{2}{5}=12\frac{4}{5}$
2. $21\frac{3}{7}-9\frac{1}{7}=12\frac{2}{7}$
3. $6\frac{1}{8}+7\frac{4}{8}=13\frac{5}{8}$
4. $1\frac{3}{4}=\frac{7}{4}$
5. $2\frac{5}{6}=\frac{17}{6}$
6. $6\frac{1}{3}=\frac{19}{3}$
7. $\frac{5}{4}=1\frac{1}{4}$
8. $\frac{23}{6}=3\frac{5}{6}$
9. $\frac{11}{4}=2\frac{3}{4}$
10. $\frac{1}{2}\text{x}\frac{3}{5}=\frac{3}{10}$
11. $\frac{12}{18}\div\frac{3}{18}=\frac{12\div3}{1}=4$
12. $\frac{4}{7}\text{x}\frac{1}{4}=\frac{4}{28}=\frac{1}{7}$
13. 7x7x7 = 343 cu. in
14. 25x25x25 = 15,625 cu. ft
15. 41x41x41 = 68,921 cu. ft
16. 2x2x2 = 8 cu. in
    8x6=48 cu. in
17. yes
18. $2\frac{1}{5}+3\frac{2}{5}=5\frac{3}{5}$ lb
19. $\frac{1}{2}\text{x}\frac{1}{4}=\frac{1}{8}$ sq. mi
20. $\frac{33}{77}>\frac{28}{77}$ so $\frac{3}{7}>\frac{4}{11}$

Lesson Practice 18A

1. done
2. done
3. $1\frac{2}{5}+4\frac{4}{5}=5\frac{6}{5}=6\frac{1}{5}$
4. $3\frac{3}{8}+1\frac{7}{8}=4\frac{10}{8}=5\frac{2}{8}=5\frac{1}{4}$
5. $7\frac{2}{5}+2\frac{4}{5}=9\frac{6}{5}=10\frac{1}{5}$

6. $2\frac{5}{6}+4\frac{3}{6}=6\frac{8}{6}=7\frac{2}{6}=7\frac{1}{3}$
7. $4\frac{2}{4}+2\frac{3}{4}=6\frac{5}{4}=7\frac{1}{4}$
8. $5\frac{5}{8}+7\frac{3}{8}=12\frac{8}{8}=13$
9. $2\frac{3}{7}+5\frac{6}{7}=7\frac{9}{7}=8\frac{2}{7}$
10. $3\frac{1}{2}+3\frac{1}{2}=6\frac{2}{2}=7$
11. $7\frac{2}{3}+3\frac{2}{3}=10\frac{4}{3}=11\frac{1}{3}$
12. $4\frac{8}{9}+8\frac{5}{9}=12\frac{13}{9}=13\frac{4}{9}$

Lesson Practice 18B

1. $9\frac{4}{6}+7\frac{5}{6}=16\frac{9}{6}=17\frac{3}{6}=17\frac{1}{2}$
2. $3\frac{4}{5}+6\frac{3}{5}=9\frac{7}{5}=10\frac{2}{5}$
3. $2\frac{7}{8}+5\frac{7}{8}=7\frac{14}{8}=8\frac{6}{8}=8\frac{3}{4}$
4. $4\frac{3}{4}+1\frac{1}{4}=5\frac{4}{4}=6$
5. $2\frac{4}{9}+3\frac{6}{9}=5\frac{10}{9}=6\frac{1}{9}$
6. $6\frac{2}{3}+8\frac{2}{3}=14\frac{4}{3}=15\frac{1}{3}$
7. $1\frac{5}{11}+2\frac{6}{11}=3\frac{11}{11}=4$
8. $7\frac{7}{10}+9\frac{5}{10}=16\frac{12}{10}=17\frac{2}{10}=17\frac{1}{5}$
9. $2\frac{4}{7}+2\frac{6}{7}=4\frac{10}{7}=5\frac{3}{7}$
10. $5\frac{1}{3}+2\frac{2}{3}=7\frac{3}{3}=8$
11. $6\frac{4}{5}+4\frac{4}{5}=10\frac{8}{5}=11\frac{3}{5}$
12. $10\frac{5}{8}+3\frac{7}{8}=13\frac{12}{8}=14\frac{4}{8}=14\frac{1}{2}$

Lesson Practice 18C

1. $4\frac{5}{8}+3\frac{3}{8}=7\frac{8}{8}=8$
2. $4\frac{5}{6}+7\frac{4}{6}=11\frac{9}{6}=12\frac{3}{6}=12\frac{1}{2}$
3. $3\frac{8}{9}+6\frac{7}{9}=9\frac{15}{9}=10\frac{6}{9}=10\frac{2}{3}$
4. $5\frac{4}{5}+2\frac{2}{5}=7\frac{6}{5}=8\frac{1}{5}$
5. $3\frac{5}{10}+4\frac{7}{10}=7\frac{12}{10}=8\frac{2}{10}=8\frac{1}{5}$
6. $7\frac{3}{4}+9\frac{2}{4}=16\frac{5}{4}=17\frac{1}{4}$
7. $2\frac{6}{12}+3\frac{7}{12}=5\frac{13}{12}=6\frac{1}{12}$
8. $8\frac{8}{11}+10\frac{6}{11}=18\frac{14}{11}=19\frac{3}{11}$
9. $6\frac{5}{8}+5\frac{7}{8}=11\frac{12}{8}=12\frac{4}{8}=12\frac{1}{2}$
10. $6\frac{2}{4}+4\frac{2}{4}=10\frac{4}{4}=11$
11. $7\frac{5}{6}+5\frac{4}{6}=12\frac{9}{6}=13\frac{3}{6}=13\frac{1}{2}$
12. $12\frac{5}{9}+8\frac{8}{9}=20\frac{13}{9}=21\frac{4}{9}$

Systematic Review18D

1. $4\frac{3}{7}+3\frac{6}{7}=7\frac{9}{7}=8\frac{2}{7}$
2. $14\frac{4}{5}+9\frac{2}{5}=23\frac{6}{5}=24\frac{1}{5}$
3. $8\frac{7}{9}+5\frac{4}{9}=13\frac{11}{9}=14\frac{2}{9}$
4. $5\frac{1}{3}=\frac{16}{3}$
5. $7\frac{2}{9}=\frac{65}{9}$
6. $3\frac{4}{5}=\frac{19}{5}$
7. $\frac{27}{5}=5\frac{2}{5}$
8. $\frac{39}{6}=6\frac{3}{6}=6\frac{1}{2}$
9. $\frac{33}{10}=3\frac{3}{10}$
10. done
11. 12x4x3 = 144 cu. in
12. 2x2x3 = 12 cu. ft
13. 12x13x8 = 1,248 cu. ft
14. $4\frac{5}{8}+3\frac{7}{8}=7\frac{12}{8}=8\frac{4}{8}=8\frac{1}{2}$ lb
15. 2x37

16. $16 \times 16 = 256$
17. $(\frac{1}{2})(10)(15) = 75$ sq. ft
18. $3{,}885 \div 7 = 555$ mi

Systematic Review18E

1. $2\frac{4}{5} + 1\frac{2}{5} = 3\frac{6}{5} = 4\frac{1}{5}$
2. $4\frac{1}{9} + 6\frac{7}{9} = 10\frac{8}{9}$
3. $5\frac{5}{6} + 2\frac{4}{6} = 7\frac{9}{6} = 8\frac{3}{6} = 8\frac{1}{2}$
4. $\frac{15}{18} - \frac{6}{18} = \frac{9}{18} = \frac{1}{2}$
5. $\frac{16}{24} - \frac{3}{24} = \frac{13}{24}$
6. $\frac{15}{20} + \frac{16}{20} = \frac{31}{20} = 1\frac{11}{20}$
7. $\frac{5}{7} \times \frac{1}{2} = \frac{5}{14}$
8. $\frac{12}{20} \div \frac{15}{20} = \frac{12 \div 15}{1} =$
   $\frac{12}{15} = \frac{4}{5}$
9. $\frac{1}{4} \times \frac{5}{6} = \frac{5}{24}$
10. $3\frac{10}{16} = 3\frac{5}{8}$
11. $12 \times 14 \times 5 = 840$ cu. in
12. $21 \times 9 \times 8 = 1{,}512$ cu. in
13. $7 \times 6 \times 10 = 420$ cu. ft
14. $20 \times 15 \times 6 = 1{,}800$ cu. ft
15. $1{,}800 \times 56 = 100{,}800$ lb
16. 16 : 2, 4, 8, 16
    20 : 2, 4, 5, 10, 20
    GCF = 4
17. $32 + 32 = 64$ ft
18. $5\frac{2}{3} + \frac{1}{3} = 5\frac{3}{3} = 6$ ft

Systematic Review18F

1. $5\frac{4}{6} + 9\frac{2}{6} = 14\frac{6}{6} = 15$
2. $11\frac{5}{8} + 3\frac{7}{8} = 14\frac{12}{8} = 15\frac{4}{8} = 15\frac{1}{2}$
3. $7\frac{7}{10} + 8\frac{9}{10} = 15\frac{16}{10} = 16\frac{6}{10} = 16\frac{3}{5}$
4. $\frac{45}{63} - \frac{14}{63} = \frac{31}{63}$
5. $\frac{5}{40} + \frac{32}{40} = \frac{37}{40}$
6. $\frac{48}{56} - \frac{7}{56} = \frac{41}{56}$
7. $\frac{40}{48} \div \frac{42}{48} = \frac{40 \div 42}{1} =$
   $\frac{40}{42} = \frac{20}{21}$
8. $\frac{1}{2} \times \frac{3}{5} = \frac{3}{10}$
9. $\frac{12}{18} \div \frac{3}{18} = \frac{12 \div 3}{1} = 4$
10. $56 \div 7 = 8$; $8 \times 4 = 32$
11. $50 \div 5 = 10$; $10 \times 3 = 30$
12. $14 \div 2 = 7$; $7 \times 1 = 7$
13. $10 \times 15 \times 7 = 1{,}050$ cu. in
14. $33 \times 19 \times 12 = 7{,}524$ cu. in
15. $5 \times 4 \times 8 = 160$ cu. ft
16. $25 \times 13 \times 8 = 2{,}600$ cu. ft full
    $2{,}600 \div 2 = 1{,}300$ cu. ft half full
17. $1{,}300 \times 56 = 72{,}800$ lb
18. yes
19. $\frac{1}{2} + \frac{1}{2} = \frac{2}{2} = 1$
    $\frac{1}{4} + \frac{1}{4} = \frac{2}{4} = \frac{1}{2}$
    $1 + \frac{1}{2} = 1\frac{1}{2}$ mi
20. $\frac{4}{5} \div \frac{4}{1} = \frac{4}{5} \div \frac{20}{5} =$
    $\frac{4 \div 20}{1} = \frac{4}{20} = \frac{1}{5}$ of a pie

Lesson Practice 19A

1. done
2. done

3. $6\frac{1}{5} - 2\frac{3}{5}$ → $5\frac{6}{5} - 2\frac{3}{5} = 3\frac{3}{5}$

4. $4\frac{1}{8} - 1\frac{5}{8}$ → $3\frac{9}{8} - 1\frac{5}{8} = 2\frac{4}{8} = 2\frac{1}{2}$

5. $2\frac{1}{4} - 1\frac{2}{4}$ → $1\frac{5}{4} - 1\frac{2}{4} = \frac{3}{4}$

6. $6 - 3\frac{7}{8}$ → $5\frac{8}{8} - 3\frac{7}{8} = 2\frac{1}{8}$

7. $7\frac{5}{16} - 2\frac{7}{16}$ → $6\frac{21}{16} - 2\frac{7}{16} = 4\frac{14}{16} = 4\frac{7}{8}$

8. $8\frac{3}{5} - 2\frac{3}{5} = 6$

9. $4\frac{1}{6} - 1\frac{5}{6}$ → $3\frac{7}{6} - 1\frac{5}{6} = 2\frac{2}{6} = 2\frac{1}{3}$

10. $9\frac{1}{8} - 2\frac{7}{8}$ → $8\frac{9}{8} - 2\frac{7}{8} = 6\frac{2}{8} = 6\frac{1}{4}$

11. $4 - \frac{1}{3}$ → $3\frac{3}{3} - \frac{1}{3} = 3\frac{2}{3}$

12. $7\frac{2}{5} - 6\frac{4}{5}$ → $6\frac{7}{5} - 6\frac{4}{5} = \frac{3}{5}$

Lesson Practice 19B

1. $14\frac{3}{7} - 2\frac{4}{7}$ → $13\frac{10}{7} - 2\frac{4}{7} = 11\frac{6}{7}$

2. $11 - 2\frac{8}{9}$ → $10\frac{9}{9} - 2\frac{8}{9} = 8\frac{1}{9}$

3. $8\frac{3}{10} - 5\frac{4}{10}$ → $7\frac{13}{10} - 5\frac{4}{10} = 2\frac{9}{10}$

4. $19\frac{2}{6} - 10\frac{3}{6}$ → $18\frac{8}{6} - 10\frac{3}{6} = 8\frac{5}{6}$

5. $3\frac{1}{3} - 1\frac{2}{3}$ → $2\frac{4}{3} - 1\frac{2}{3} = 1\frac{2}{3}$

6. $\begin{array}{r} 5 \\ -1\frac{4}{7} \\ \hline \end{array}$ $\quad$ $\begin{array}{r} 4\frac{7}{7} \\ -1\frac{4}{7} \\ \hline 3\frac{3}{7} \end{array}$

7. $\begin{array}{r} 5\frac{1}{4} \\ -2\frac{3}{4} \\ \hline \end{array}$ $\quad$ $\begin{array}{r} 4\frac{5}{4} \\ -2\frac{3}{4} \\ \hline 2\frac{2}{4} \end{array} = 2\frac{1}{2}$

8. $\begin{array}{r} 6\frac{1}{5} \\ -2\frac{3}{5} \\ \hline \end{array}$ $\quad$ $\begin{array}{r} 5\frac{6}{5} \\ -2\frac{3}{5} \\ \hline 3\frac{3}{5} \end{array}$

9. $\begin{array}{r} 4\frac{1}{8} \\ -1\frac{5}{8} \\ \hline \end{array}$ $\quad$ $\begin{array}{r} 3\frac{9}{8} \\ -1\frac{5}{8} \\ \hline 2\frac{4}{8} \end{array} = 2\frac{1}{2}$

10. $\begin{array}{r} 6\frac{5}{8} \\ -3\frac{7}{8} \\ \hline \end{array}$ $\quad$ $\begin{array}{r} 5\frac{13}{8} \\ -3\frac{7}{8} \\ \hline 2\frac{6}{8} \end{array} = 2\frac{3}{4}$

11. $\begin{array}{r} 8 \\ -\frac{3}{10} \\ \hline \end{array}$ $\quad$ $\begin{array}{r} 7\frac{10}{10} \\ -\frac{3}{10} \\ \hline 7\frac{7}{10} \end{array}$

12. $\begin{array}{r} 2\frac{1}{4} \\ -1\frac{3}{4} \\ \hline \end{array}$ $\quad$ $\begin{array}{r} 1\frac{5}{4} \\ -1\frac{3}{4} \\ \hline \frac{2}{4} \end{array} = \frac{1}{2}$

Lesson Practice 19C

1. $\begin{array}{r} 5\frac{1}{3} \\ -2\frac{2}{3} \\ \hline \end{array}$ $\quad$ $\begin{array}{r} 4\frac{4}{3} \\ -2\frac{2}{3} \\ \hline 2\frac{2}{3} \end{array}$

2. $\begin{array}{r} 6 \\ -2\frac{3}{8} \\ \hline \end{array}$ $\quad$ $\begin{array}{r} 5\frac{8}{8} \\ -2\frac{3}{8} \\ \hline 3\frac{5}{8} \end{array}$

3. $\begin{array}{r} 6\frac{3}{16} \\ -2\frac{5}{16} \\ \hline \end{array}$ $\quad$ $\begin{array}{r} 5\frac{19}{16} \\ -2\frac{5}{16} \\ \hline 3\frac{14}{16} \end{array} = 3\frac{7}{8}$

4. $\begin{array}{r} 8\frac{1}{5} \\ -2\frac{4}{5} \\ \hline \end{array}$ $\quad$ $\begin{array}{r} 7\frac{6}{5} \\ -2\frac{4}{5} \\ \hline 5\frac{2}{5} \end{array}$

5. $\begin{array}{r} 9\frac{1}{4} \\ -6\frac{3}{4} \\ \hline \end{array}$ $\quad$ $\begin{array}{r} 8\frac{5}{4} \\ -6\frac{3}{4} \\ \hline 2\frac{2}{4} \end{array} = 2\frac{1}{2}$

6. $\begin{array}{r} 9 \\ -1\frac{3}{5} \\ \hline \end{array}$ $\quad$ $\begin{array}{r} 8\frac{5}{5} \\ -1\frac{3}{5} \\ \hline 7\frac{2}{5} \end{array}$

7. $\begin{array}{r} 4\frac{1}{6} \\ -1\frac{1}{6} \\ \hline 3 \end{array}$

8. $\begin{array}{r} 5\frac{5}{8} \\ -2\frac{7}{8} \\ \hline \end{array}$ $\quad$ $\begin{array}{r} 4\frac{13}{8} \\ -2\frac{7}{8} \\ \hline 2\frac{6}{8} \end{array} = 2\frac{3}{4}$

9. $6\frac{1}{3} - 2\frac{2}{3}$ → $5\frac{4}{3} - 2\frac{2}{3} = 3\frac{2}{3}$

10. $18\frac{1}{4} - 7\frac{2}{4}$ → $17\frac{5}{4} - 7\frac{2}{4} = 10\frac{3}{4}$

11. $6 - \frac{4}{5}$ → $5\frac{5}{5} - \frac{4}{5} = 5\frac{1}{5}$

12. $9\frac{1}{6} - 2\frac{5}{6}$ → $8\frac{7}{6} - 2\frac{5}{6} = 6\frac{2}{6} = 6\frac{1}{3}$

Systematic Review19D

1. $5\frac{1}{5} - 2\frac{3}{5}$ → $4\frac{6}{5} - 2\frac{3}{5} = 2\frac{3}{5}$

2. $8\frac{5}{8} - 2\frac{7}{8}$ → $7\frac{13}{8} - 2\frac{7}{8} = 5\frac{6}{8} = 5\frac{3}{4}$

3. $6 - 4\frac{2}{3}$ → $5\frac{3}{3} - 4\frac{2}{3} = 1\frac{1}{3}$

4. $7\frac{1}{5} + 2\frac{4}{5} = 9\frac{5}{5} = 10$

5. $1\frac{3}{4} + 2\frac{3}{4} = 3\frac{6}{4} = 4\frac{2}{4} = 4\frac{1}{2}$

6. $9\frac{7}{12} + 3\frac{11}{12} = 12\frac{18}{12} = 13\frac{6}{12} = 13\frac{1}{2}$

7. $\frac{1}{6} \times \frac{4}{5} = \frac{4}{30} = \frac{2}{15}$

8. $\frac{3}{6} \times \frac{1}{2} = \frac{3}{12} = \frac{1}{4}$

9. $\frac{4}{5} \times \frac{2}{7} = \frac{8}{35}$

10. done

11. done

12. $6 \div 3 = 2$ yd

13. $5 \times 3 = 15$ ft

14. $3 \times 3 = 9$ ft
    $9 \times 15 = 135$ sq. ft

15. $15 \div 3 = 5$ yd
    $5 \times 3 = 15$ sq. yd

16. $\frac{3}{8} + \frac{1}{8} = \frac{4}{8}$
    $\frac{4}{8} + \frac{3}{4} = \frac{16}{32} + \frac{24}{32} =$
    $\frac{40}{32} = 1\frac{8}{32} = 1\frac{1}{4}$ ft

17. $4\frac{5}{4} - 3\frac{3}{4} = 1\frac{2}{4} = 1\frac{1}{2}$ ft

18. $3 \div 3 = 1$ yd
    $1 \times 1 \times 1 = 1$ cu. yd

Systematic Review19E

1. $7\frac{3}{6} - 2\frac{5}{6}$ → $6\frac{9}{6} - 2\frac{5}{6} = 4\frac{4}{6} = 4\frac{2}{3}$
2. $6\frac{4}{7} - \frac{6}{7}$ → $5\frac{11}{7} - \frac{6}{7} = 5\frac{5}{7}$
3. $5 - 4\frac{6}{8}$ → $4\frac{8}{8} - 4\frac{6}{8} = \frac{2}{8} = \frac{1}{4}$
4. $2\frac{1}{2} + 3\frac{1}{2} = 5\frac{2}{2} = 6$
5. $8\frac{2}{7} + 8\frac{6}{7} = 16\frac{8}{7} = 17\frac{1}{7}$
6. $5\frac{4}{7} + 2\frac{5}{7} = 7\frac{9}{7} = 8\frac{2}{7}$
7. $\frac{7}{8} \div \frac{1}{8} = \frac{7 \div 1}{1} = 7$
8. $\frac{15}{27} \div \frac{18}{27} = \frac{15 \div 18}{1} = \frac{15}{18} = \frac{5}{6}$
9. $\frac{4}{12} \div \frac{9}{12} = \frac{4 \div 9}{1} = \frac{4}{9}$
10. $1\frac{2}{10} = 1\frac{1}{5}$
11. $24 \div 3 = 8$ yd
12. $10 \text{x} 3 = 30$ ft
13. $36 \div 3 = 12$ yd
14. $16 \text{x} 3 = 48$ ft
15. $\frac{1}{2} \text{x} \frac{2}{5} = \frac{2}{10} = \frac{1}{5}$

    $\frac{1}{5} \text{x} \frac{1}{2} = \frac{1}{10}$ sq. mi
16. $(\frac{1}{2})(20)(4) = 40$ cu. ft
17. $3\frac{1}{8} + 4\frac{5}{8} = 7\frac{6}{8}$

    $7\frac{6}{8} - 2\frac{7}{8} =$

    $6\frac{14}{8} - 2\frac{7}{8} = 4\frac{7}{8}$ lb
18. $125 \div 25 = 5$ ft

Systematic Review19F

1. $25\frac{1}{3} - 12\frac{1}{3} = 13$
2. $1\frac{5}{8} - \frac{6}{8}$ → $\frac{13}{8} - \frac{6}{8} = \frac{7}{8}$
3. $9 - 4\frac{2}{7}$ → $8\frac{7}{7} - 4\frac{2}{7} = 4\frac{5}{7}$
4. $1\frac{3}{4} + 2\frac{1}{4} = 3\frac{4}{4} = 4$
5. $10\frac{2}{6} + 7\frac{5}{6} = 17\frac{7}{6} = 18\frac{1}{6}$

6. $\begin{array}{r} 9\frac{5}{9} \\ +5\frac{8}{9} \\ \hline 14\frac{13}{9} = 15\frac{4}{9} \end{array}$

7. $\frac{7}{14} + \frac{6}{14} = \frac{13}{14}$

$\frac{13}{14} + \frac{4}{9} = \frac{117}{126} + \frac{56}{126} = \frac{173}{126} = 1\frac{47}{126}$

8. $\frac{3}{6} + \frac{4}{6} = \frac{7}{6}; \ \frac{7}{6} + \frac{3}{5} = \frac{35}{30} + \frac{18}{30} = \frac{53}{30} = 1\frac{23}{30}$

9. $\frac{8}{12} + \frac{3}{12} + \frac{1}{12} = \frac{12}{12} = 1$

10. $\frac{15}{20} < \frac{16}{20}$

11. $\frac{32}{72} > \frac{27}{72}$

12. $\frac{72}{132} > \frac{55}{132}$

13. $45 \div 3 = 15$ yd

14. $17 \text{x} 3 = 51$ ft

15. $3 \text{x} 3 = 9$ ft

16. $\frac{5}{8} \text{x} \frac{1}{5} = \frac{5}{40} = \frac{1}{8}$ sq. mi

17. $\frac{5}{8} + \frac{5}{8} = \frac{10}{8} = \frac{5}{4}; \ \frac{1}{5} + \frac{1}{5} = \frac{2}{5}$

$\frac{5}{4} + \frac{2}{5} = \frac{25}{20} + \frac{8}{20} = \frac{33}{20} = 1\frac{13}{20}$ mi

18. $2\frac{3}{5} - 1\frac{4}{5} = 1\frac{8}{5} - 1\frac{4}{5} = \frac{4}{5}$ bushel

19. $30 \text{x} 3 = 90$ ft

20. $1{,}260 \div 45 = 28$ ft

Lesson Practice 20A

1. done
2. done
3. $\begin{array}{r} 4\frac{1}{3} + \frac{1}{3} = 4\frac{2}{3} \\ -\ 3\frac{2}{3} + \frac{1}{3} = 4 \\ \hline \frac{2}{3} \end{array}$

4. $\begin{array}{r} 7 + \frac{1}{4} = 7\frac{1}{4} \\ -\ 2\frac{3}{4} + \frac{1}{4} = 3 \\ \hline 4\frac{1}{4} \end{array}$

5. $\begin{array}{r} 6\frac{1}{5} + \frac{3}{5} = 6\frac{4}{5} \\ -\ 3\frac{2}{5} + \frac{3}{5} = 4 \\ \hline 2\frac{4}{5} \end{array}$

6. $\begin{array}{r} 8 + \frac{3}{8} = 8\frac{3}{8} \\ -1\frac{5}{8} + \frac{3}{8} = 2 \\ \hline 6\frac{3}{8} \end{array}$

7. $\begin{array}{r} 9\frac{5}{16} + \frac{9}{16} = 9\frac{14}{16} \\ -\ 2\frac{7}{16} + \frac{9}{16} = 3 \\ \hline 6\frac{14}{16} = 6\frac{7}{8} \end{array}$

8. $\begin{array}{r} 10 + \frac{1}{8} = 10\frac{1}{8} \\ -\ 4\frac{7}{8} + \frac{1}{8} = 5 \\ \hline 5\frac{1}{8} \end{array}$

Lesson Practice 20B

1. $\begin{array}{r} 5\frac{2}{5} + \frac{1}{5} = 5\frac{3}{5} \\ -\ 1\frac{4}{5} + \frac{1}{5} = 2 \\ \hline 3\frac{3}{5} \end{array}$

2. $\begin{array}{r} 9 + \frac{1}{4} = 9\frac{1}{4} \\ -\ 2\frac{3}{4} + \frac{1}{4} = 3 \\ \hline 6\frac{1}{4} \end{array}$

3.
$$\begin{array}{rcccl} & 10\frac{1}{8} & + \frac{1}{8} & = & 10\frac{2}{8} \\ - & 2\frac{7}{8} & + \frac{1}{8} & = & 3 \\ \hline & & & & 7\frac{2}{8} = 7\frac{1}{4} \end{array}$$

4.
$$\begin{array}{rl} & 7 + \frac{2}{3} = 7\frac{2}{3} \\ - & 6\frac{1}{3} + \frac{2}{3} = 7 \\ \hline & \frac{2}{3} \end{array}$$

5.
$$\begin{array}{rl} & 12\frac{1}{6} + \frac{4}{6} = 12\frac{5}{6} \\ - & 8\frac{2}{6} + \frac{4}{6} = 9 \\ \hline & 3\frac{5}{6} \end{array}$$

6.
$$\begin{array}{rl} & 25 + \frac{2}{2} = 25\frac{2}{5} \\ - & 5\frac{3}{5} + \frac{2}{5} = 6 \\ \hline & 19\frac{2}{5} \end{array}$$

7.
$$\begin{array}{rcccl} & 8\frac{3}{10} & + \frac{1}{10} & = & 8\frac{4}{10} \\ - & 1\frac{9}{10} & + \frac{1}{10} & = & 2 \\ \hline & & & & 6\frac{4}{10} = 6\frac{2}{5} \end{array}$$

8.
$$\begin{array}{rl} & 4 + \frac{3}{8} = 4\frac{3}{8} \\ - & 3\frac{5}{8} + \frac{3}{8} = 4 \\ \hline & \frac{3}{8} \end{array}$$

Lesson Practice 20C

1.
$$\begin{array}{rcccl} & 6\frac{3}{6} & + \frac{1}{6} & = & 6\frac{4}{6} \\ - & 2\frac{5}{6} & + \frac{1}{6} & = & 3 \\ \hline & & & & 3\frac{4}{6} = 3\frac{2}{3} \end{array}$$

2.
$$\begin{array}{rl} & 14 + \frac{1}{5} = 14\frac{1}{5} \\ - & 5\frac{4}{5} + \frac{1}{5} = 6 \\ \hline & 8\frac{1}{5} \end{array}$$

3.
$$\begin{array}{rl} & 16\frac{2}{9} + \frac{5}{9} = 16\frac{7}{9} \\ - & 7\frac{4}{9} + \frac{5}{9} = 8 \\ \hline & 8\frac{7}{9} \end{array}$$

4.
$$\begin{array}{rl} & 9 + \frac{1}{4} = 9\frac{1}{4} \\ - & 3\frac{3}{4} + \frac{1}{4} = 4 \\ \hline & 5\frac{1}{4} \end{array}$$

5.
$$\begin{array}{rl} & 13\frac{3}{7} + \frac{1}{7} = 13\frac{4}{7} \\ - & 8\frac{6}{7} + \frac{1}{7} = 9 \\ \hline & 4\frac{4}{7} \end{array}$$

6.
$$\begin{array}{rl} & 30 + \frac{3}{10} = 30\frac{3}{10} \\ - & 11\frac{7}{10} + \frac{3}{10} = 12 \\ \hline & 18\frac{3}{10} \end{array}$$

7.
$$\begin{array}{rl} & 6\frac{1}{4} + \frac{2}{4} = 6\frac{3}{4} \\ - & 2\frac{2}{4} + \frac{2}{4} = 3 \\ \hline & 3\frac{3}{4} \end{array}$$

8.
$$\begin{array}{rl} & 5 + \frac{5}{8} = 5\frac{5}{8} \\ - & 1\frac{3}{8} + \frac{5}{8} = 2 \\ \hline & 3\frac{5}{8} \end{array}$$

Systematic Review 20D

1.
$$5\frac{2}{7}+\frac{1}{7}=5\frac{3}{7}$$
$$-\ 3\frac{6}{7}+\frac{1}{7}=4$$
$$1\frac{3}{7}$$

2.
$$19+\frac{1}{3}=19\frac{1}{3}$$
$$-\ 6\frac{2}{3}+\frac{1}{3}=7$$
$$12\frac{1}{3}$$

3. $2\frac{5}{8}+1\frac{7}{8}=3\frac{12}{8}=4\frac{4}{8}=4\frac{1}{2}$
4. $3\frac{7}{5}-1\frac{4}{5}=2\frac{3}{5}$
5. $3\frac{1}{2}+5\frac{1}{2}=8\frac{2}{2}=9$
6. 5x5
7. 2x2x3x3
8. 2x2x11
9. $\frac{3}{7}=\frac{6}{14}=\frac{9}{21}$
10. $\frac{2}{3}=\frac{4}{6}=\frac{6}{9}$
11. $\frac{1}{9}=\frac{2}{18}=\frac{3}{27}$
12. done
13. done
14. $12\div2=6$ qt
15. $11\times2=22$ pt
16. $19\frac{2}{2}-5\frac{1}{2}=14\frac{1}{2}$ lb
17. $7\times2=14$ pint jars
18. $6\times3=18$ ft
    18x18x18=5,832 cu. ft

Systematic Review 20E

1.
$$7\frac{4}{9}\ +\ \frac{2}{9}\ =\ 7\frac{6}{9}$$
$$-\ 3\frac{7}{9}\ +\ \frac{2}{9}\ =\ 4$$
$$3\frac{6}{9}\ =\ 3\frac{2}{3}$$

2.
$$13+\frac{7}{8}=13\frac{7}{8}$$
$$-\ 7\frac{1}{8}+\frac{7}{8}=8$$
$$5\frac{7}{8}$$

3. $2\frac{1}{5}+4\frac{3}{5}=6\frac{4}{5}$
4. $5\frac{9}{6}-4\frac{5}{6}=1\frac{4}{6}=1\frac{2}{3}$
5. $5\frac{2}{3}+5\frac{2}{3}=10\frac{4}{3}=11\frac{1}{3}$
6. 25: 5, 25
   35: 5, 7, 35
   GCF = 5
7. 12: 2, 3, 4, 6, 12
   36: 2, 3, 4, 6, 9, 12, 18, 36
   GCF = 12
8. 42: 2, 3, 6, 7, 14, 21, 42
   49: 7, 49
   GCF = 7
9. $\frac{6}{12}-\frac{2}{12}=\frac{4}{12}=\frac{1}{3}$
10. $\frac{35}{50}-\frac{20}{50}=\frac{15}{50}=\frac{3}{10}$
11. $\frac{99}{108}-\frac{60}{108}=\frac{39}{108}=\frac{13}{36}$
12. $11\times2=22$ pt
13. $18\div2=9$ qt
14. $22\div2=11$ qt
15. $2\frac{5}{8}+3\frac{7}{8}=5\frac{12}{8}=6\frac{4}{8}=6\frac{1}{2}$ in
16. no
17. $16\div2=8$ qt
18. Answers will vary.
19. $25+43=68$
    $\frac{1}{4}\times68=\$17$
20. $7\times7=49$

Systematic Review 20F

1. $$\begin{array}{r} 9\frac{1}{6} + \frac{1}{6} = 9\frac{2}{6} \\ -\ 4\frac{5}{6} + \frac{1}{6} = 5 \\ \hline 4\frac{2}{6} = 4\frac{1}{3} \end{array}$$
2. $$\begin{array}{r} 18 + \frac{1}{2} = 18\frac{1}{2} \\ -\ 12\frac{1}{2} + \frac{1}{2} = 13 \\ \hline 5\frac{1}{2} \end{array}$$
3. $3\frac{3}{8} + 5\frac{5}{8} = 8\frac{8}{8} = 9$
4. $8\frac{1}{3} - 6\frac{1}{3} = 2$
5. $7\frac{5}{9} + 7\frac{8}{9} = 14\frac{13}{9} = 15\frac{4}{9}$
6. $192 \div 3 = 64$; $64 \text{x} 2 = 128$
7. $555 \div 5 = 111$; $111 \text{x} 1 = 111$
8. $84 \div 4 = 21$; $21 \text{x} 3 = 63$
9. $\frac{6}{8} + \frac{4}{8} = \frac{10}{8} = \frac{5}{4} = 1\frac{1}{4}$
10. $\frac{40}{110} + \frac{33}{110} = \frac{73}{110}$
11. $\frac{15}{27} + \frac{9}{27} = \frac{24}{27} = \frac{8}{9}$
12. $12 \text{x} 2 = 24$ pt
13. $28 \div 2 = 14$ qt
14. $15 \text{x} 3 = 45$ ft
15. $4\frac{2}{10} - 1\frac{9}{10} = 3\frac{12}{10} - 1\frac{9}{10} = 2\frac{3}{10}$ in
16. yes
17. $4 + 6 = 10$ qt of juice
    $10 \div 5 = 2$ qt per person
    $2 \text{x} 2 = 4$ pt per person
18. 11 in
19. $1\frac{1}{2} + 1\frac{1}{2} + 1\frac{1}{2} + 1\frac{1}{2} = 4\frac{4}{2} = 6$ ft
20. $\frac{1}{8} \div \frac{1}{4} = \frac{4}{32} \div \frac{8}{32} = \frac{4 \div 8}{1} = \frac{4}{8} = \frac{1}{2}$ mi

Lesson Practice 21A

1. done
2. $3\frac{15}{40} + 1\frac{32}{40} = 4\frac{47}{40} = 5\frac{7}{40}$
3. $18\frac{32}{80} + 3\frac{50}{80} = 21\frac{82}{80} = 22\frac{2}{80} = 22\frac{1}{40}$
4. $11\frac{18}{30} + 4\frac{25}{30} = 15\frac{43}{30} = 16\frac{13}{30}$
5. $9\frac{4}{12} + 6\frac{3}{12} = 15\frac{7}{12}$
6. $4\frac{10}{15} + 1\frac{6}{15} = 5\frac{16}{15} = 6\frac{1}{15}$
7. $9\frac{30}{50} + 2\frac{35}{50} = 11\frac{65}{50} = 12\frac{15}{50} = 12\frac{3}{10}$
8. $12\frac{72}{80} + 4\frac{50}{80} = 16\frac{122}{80} = 16\frac{61}{40} = 17\frac{21}{40}$
9. $3\frac{4}{8} + 1\frac{6}{8} = 4\frac{10}{8} = 5\frac{2}{8} = 5\frac{1}{4}$
10. $4\frac{20}{32} + 1\frac{8}{32} = 5\frac{28}{32} = 5\frac{7}{8}$

Lesson Practice 21B

1. $7\frac{15}{20} + 9\frac{12}{20} = 16\frac{27}{20} = 17\frac{7}{20}$
2. $6\frac{18}{24} + 6\frac{20}{24} = 12\frac{38}{24} = 12\frac{19}{12} = 13\frac{7}{12}$
3. $19\frac{70}{80} + 2\frac{40}{80} = 21\frac{110}{80} = 21\frac{11}{8} = 22\frac{3}{8}$
4. $12\frac{24}{33} + 5\frac{22}{33} = 17\frac{46}{33} = 18\frac{13}{33}$
5. $8\frac{12}{15} + 2\frac{5}{15} = 10\frac{17}{15} = 11\frac{2}{15}$
6. $4\frac{15}{20} + 2\frac{4}{20} = 6\frac{19}{20}$
7. $6\frac{15}{27} + 8\frac{9}{27} = 14\frac{24}{27} = 14\frac{8}{9}$
8. $6\frac{18}{27} + 1\frac{21}{27} = 7\frac{39}{27} = 7\frac{13}{9} = 8\frac{4}{9}$
9. $7\frac{15}{50} + 9\frac{40}{50} = 16\frac{55}{50} = 16\frac{11}{10} = 17\frac{1}{10}$
10. $2\frac{6}{10} + 1\frac{5}{10} = 3\frac{11}{10} = 4\frac{1}{10}$

Lesson Practice 21C

1. $6\frac{4}{10}+2\frac{5}{10}=8\frac{9}{10}$
2. $5\frac{27}{63}+3\frac{7}{63}=8\frac{34}{63}$
3. $2\frac{5}{40}+\frac{32}{40}=2\frac{37}{40}$
4. $8\frac{27}{63}+7\frac{14}{63}=15\frac{41}{63}$
5. $3\frac{8}{10}+4\frac{5}{10}=7\frac{13}{10}=8\frac{3}{10}$
6. $2\frac{7}{21}+5\frac{15}{21}=7\frac{22}{21}=8\frac{1}{21}$
7. $6\frac{6}{10}+4\frac{5}{10}=10\frac{11}{10}=11\frac{1}{10}$
8. $1\frac{15}{20}+5\frac{12}{20}=6\frac{27}{20}=7\frac{7}{20}$
9. $13\frac{12}{42}+4\frac{7}{42}=17\frac{19}{42}$
10. $16\frac{4}{8}+17\frac{6}{8}=33\frac{10}{8}=34\frac{2}{8}=34\frac{1}{4}$

Systematic Review 21D

1. $8\frac{25}{35}+4\frac{21}{35}=12\frac{46}{35}=13\frac{11}{35}$
2. $6\frac{20}{28}+1\frac{21}{28}=7\frac{41}{28}=8\frac{13}{28}$
3. $4\frac{7}{6}-2\frac{4}{6}=2\frac{3}{6}=2\frac{1}{2}$
4. $6\frac{2}{5}-1\frac{1}{5}=5\frac{1}{5}$
5. $13\frac{13}{10}-9\frac{9}{10}=4\frac{4}{10}=4\frac{2}{5}$
6. $\frac{2}{6}\div\frac{3}{6}=\frac{2\div3}{1}=\frac{2}{3}$
7. $\frac{9}{18}\div\frac{12}{18}=\frac{9\div12}{1}=\frac{9}{12}=\frac{3}{4}$
8. $\frac{20}{50}\div\frac{20}{50}=\frac{20\div20}{1}=1$
9. $\frac{1}{8}\text{x}\frac{6}{7}=\frac{6}{56}=\frac{3}{28}$
10. $\frac{4}{5}\text{x}\frac{1}{9}=\frac{4}{45}$
11. $\frac{3}{5}\text{x}\frac{5}{6}=\frac{15}{30}=\frac{1}{2}$
12. $99\div3=33$ yd
13. $6\div2=3$ qt
14. $36\text{x}2=72$ pt
15. $5+4=9$
    $9\times8=72$
    $72-2=70$
    $70\times10=7$
    $7+3=10$
16. $7-3=4$
    $4\times6=24$
    $24\div3=8$
    $8\times9=72$
17. $13\div3=4\frac{1}{3}$ yd
18. $1\frac{4}{8}+1\frac{6}{8}=2\frac{10}{8}=2\frac{5}{4}=3\frac{1}{4}$ hr

Systematic Review 21E

1. $8\frac{8}{12}+5\frac{9}{12}=13\frac{17}{12}=14\frac{5}{12}$
2. $3\frac{15}{40}+1\frac{16}{40}=4\frac{31}{40}$
3. $2\frac{5}{4}-1\frac{3}{4}=1\frac{2}{4}=1\frac{1}{2}$
4. $6\frac{6}{5}-6\frac{4}{5}=\frac{2}{5}$
5. $4\frac{3}{3}-1\frac{2}{3}=3\frac{1}{3}$
6. $\frac{20}{32}\div\frac{8}{32}=\frac{20\div8}{1}=\frac{20}{8}=\frac{5}{2}=2\frac{1}{2}$
7. $\frac{3}{7}\div\frac{1}{7}=\frac{3\div1}{1}=3$
8. $\frac{15}{27}\div\frac{18}{27}=\frac{15\div18}{1}=\frac{15}{18}=\frac{5}{6}$
9. $\frac{1}{9}\text{x}\frac{7}{8}=\frac{7}{72}$
10. $\frac{2}{3}\text{x}\frac{1}{2}=\frac{2}{6}=\frac{1}{3}$
11. $\frac{7}{10}\text{x}\frac{1}{5}=\frac{7}{50}$
12. $7\frac{1}{2}+7\frac{1}{2}=14\frac{2}{2}=15$
    $3\frac{1}{4}+3\frac{1}{4}=6\frac{2}{4}=6\frac{1}{2}$
    $15+6\frac{1}{2}=21\frac{1}{2}$ in
13. $5\frac{1}{2}+5\frac{1}{2}+5\frac{1}{2}+5\frac{1}{2}=20\frac{4}{2}=$ 22 in

14. $4\frac{1}{3}+5\frac{2}{3}=9\frac{3}{3}=10$

    $10+3\frac{1}{2}=13\frac{1}{2}$ ft
15. $8x6x3=144$ cu. ft
16. $144\div12=12$ cu. ft drained
    144-12=132 cu. ft left
17. $17{,}259\div3=5{,}753$ yd
18. $25\div2=12\frac{1}{2}$ qt
19. Answers will vary.
20. $21\div7=3$
    $3x4=12$
    $12+3=15$
    $15-10=5$

15. $\frac{5}{6}x\frac{3}{5}=\frac{15}{30}=\frac{1}{2}$ of the job
16. $5x5x5=125$ cu. ft
    125x56=7,000 lb
17. $577\div3=192\frac{1}{3}$ yd
18. $10-5\frac{1}{2}=9\frac{2}{2}-5\frac{1}{2}=4\frac{1}{2}$ years
19. Answers will vary.
20. $8x7=56$
    $56-1=55$
    $55\div5=11$
    $11-3=8$

Systematic Review 21F

1. $18\frac{56}{80}+3\frac{50}{80}=21\frac{106}{80}=22\frac{26}{80}=22\frac{13}{40}$
2. $11\frac{18}{30}+4\frac{5}{30}=15\frac{23}{30}$
3. $9\frac{3}{7}-5\frac{1}{7}=4\frac{2}{7}$
4. $16\frac{7}{5}-8\frac{3}{5}=8\frac{4}{5}$
5. $5\frac{8}{8}-2\frac{7}{8}=3\frac{1}{8}$
6. $\frac{24}{32}\div\frac{8}{32}=\frac{24\div8}{1}=3$
7. $\frac{40}{50}\div\frac{5}{50}=\frac{40\div5}{1}=8$
8. $\frac{60}{72}\div\frac{6}{72}=\frac{60\div6}{1}=10$
9. $\frac{4}{5}x\frac{1}{6}=\frac{4}{30}=\frac{2}{15}$
10. $\frac{5}{6}x\frac{2}{8}=\frac{10}{48}=\frac{5}{24}$
11. $\frac{3}{5}x\frac{4}{7}=\frac{12}{35}$
12. $\frac{2}{3}x\frac{1}{3}=\frac{2}{9}$ sq. ft
13. $\frac{3}{4}x\frac{3}{4}=\frac{9}{16}$ sq. ft
14. $\frac{1}{2}x2x\frac{5}{7}=\frac{5}{7}$ sq. in

Lesson Practice 22A

1. done
2. done
3. $9\frac{8}{24}-6\frac{9}{24}=8\frac{32}{24}-6\frac{9}{24}=2\frac{23}{24}$
4. $10\frac{15}{18}-3\frac{6}{18}=7\frac{9}{18}=7\frac{1}{2}$
5. $7\frac{27}{45}-2\frac{40}{45}=6\frac{72}{45}-2\frac{40}{45}=4\frac{32}{45}$
6. $9\frac{2}{16}-8\frac{8}{16}=8\frac{18}{16}-8\frac{8}{16}=\frac{10}{16}=\frac{5}{8}$
7. $8\frac{4}{8}-2\frac{6}{8}=7\frac{12}{8}-2\frac{6}{8}=5\frac{6}{8}=5\frac{3}{4}$
8. $5\frac{6}{18}-2\frac{15}{18}=4\frac{24}{18}-2\frac{15}{18}=2\frac{9}{18}=2\frac{1}{2}$
9. $11\frac{4}{12}-3\frac{3}{12}=8\frac{1}{12}$
10. $26\frac{10}{15}-22\frac{12}{15}=25\frac{25}{15}-22\frac{12}{15}=3\frac{13}{15}$

Lesson Practice 22B

1. $10\frac{7}{28}-3\frac{16}{28}=9\frac{35}{28}-3\frac{16}{28}=6\frac{19}{28}$
2. $9\frac{6}{18}-2\frac{15}{18}=8\frac{24}{18}-2\frac{15}{18}=6\frac{9}{18}=6\frac{1}{2}$
3. $7\frac{8}{10}-7\frac{5}{10}=\frac{3}{10}$
4. $3\frac{5}{15}-1\frac{12}{15}=2\frac{20}{15}-1\frac{12}{15}=1\frac{8}{15}$

5. $5\frac{5}{30}-2\frac{24}{30}=4\frac{35}{30}-2\frac{24}{30}=2\frac{11}{30}$
6. $8\frac{6}{10}-4\frac{5}{10}=4\frac{1}{10}$
7. $15\frac{20}{28}-12\frac{21}{28}=14\frac{48}{28}-12\frac{21}{28}=2\frac{27}{28}$
8. $2\frac{4}{8}-2\frac{2}{8}=\frac{2}{8}=\frac{1}{4}$
9. $7\frac{9}{30}-5\frac{20}{30}=6\frac{39}{30}-5\frac{20}{30}=1\frac{19}{30}$
10. $8\frac{9}{18}-4\frac{4}{18}=4\frac{5}{18}$

Lesson Practice 22C

1. $8\frac{10}{60}-1\frac{18}{60}=7\frac{70}{60}-1\frac{18}{60}=6\frac{52}{60}=6\frac{13}{15}$
2. $5\frac{6}{24}-2\frac{20}{24}=4\frac{30}{24}-2\frac{20}{24}=2\frac{10}{24}=2\frac{5}{12}$
3. $14\frac{24}{32}-11\frac{28}{32}=13\frac{56}{32}-11\frac{28}{32}=2\frac{28}{32}=2\frac{7}{8}$
4. $7\frac{8}{20}-3\frac{5}{20}=4\frac{3}{20}$
5. $8\frac{3}{18}-2\frac{12}{18}=7\frac{21}{18}-2\frac{12}{18}=5\frac{9}{18}=5\frac{1}{2}$
6. $10\frac{5}{10}-3\frac{8}{10}=9\frac{15}{10}-3\frac{8}{10}=6\frac{7}{10}$
7. $8\frac{10}{14}-2\frac{7}{14}=6\frac{3}{14}$
8. $4\frac{4}{8}-1\frac{6}{8}=3\frac{12}{8}-1\frac{6}{8}=2\frac{6}{8}=2\frac{3}{4}$
9. $6\frac{12}{21}-4\frac{14}{21}=5\frac{33}{21}-4\frac{14}{21}=1\frac{19}{21}$
10. $10\frac{5}{30}-5\frac{12}{30}=9\frac{35}{30}-5\frac{12}{30}=4\frac{23}{30}$

Systematic Review 22D

1. $8\frac{8}{28}-2\frac{21}{28}=7\frac{36}{28}-2\frac{21}{28}=5\frac{15}{28}$
2. $14\frac{2}{6}-3\frac{3}{6}=13\frac{8}{6}-3\frac{3}{6}=10\frac{5}{6}$
3. $6\frac{12}{15}+1\frac{10}{15}=7\frac{22}{15}=8\frac{7}{15}$
4. $4\frac{3}{5}+1\frac{2}{5}=5\frac{5}{5}=6$
5. $12\frac{7}{56}+9\frac{32}{56}=21\frac{39}{56}$
6. $\frac{1}{2}\div\frac{1}{2}=\frac{1\div1}{1}=1$
7. $\frac{1}{4}\text{x}\frac{5}{6}=\frac{5}{24}$
8. $\frac{40}{48}\div\frac{42}{48}=\frac{40\div42}{1}=\frac{40}{42}=\frac{20}{21}$
9. $6\frac{2}{3}=\frac{20}{3}$
10. $27\frac{4}{9}=\frac{247}{9}$
11. $13\frac{3}{5}=\frac{68}{5}$
12. done
13. done
14. $10\text{x}4=40$ qt
15. $11\div4=2\frac{3}{4}$ gal
16. $4\frac{3}{10}-3\frac{9}{10}=3\frac{13}{10}-3\frac{9}{10}=$
    $\frac{4}{10}=\frac{2}{5}$ yd
17. $2\frac{5}{8}$"x$6\frac{1}{8}$"
    (Answers that are close may be accepted.)
18. $9-1=8$
    $8\text{x}7=56$
    $56+4=60$
    $60\div6=10$

Systematic Review 22E

1. $10\frac{7}{56}-4\frac{48}{56}=9\frac{63}{56}-4\frac{48}{56}=5\frac{15}{56}$
2. $4\frac{4}{20}-3\frac{15}{20}=3\frac{24}{20}-3\frac{15}{20}=\frac{9}{20}$
3. $2\frac{5}{50}+6\frac{10}{50}=8\frac{15}{50}=8\frac{3}{10}$
4. $2\frac{21}{24}+1\frac{16}{24}=3\frac{37}{24}=4\frac{13}{24}$
5. $25\frac{16}{72}+7\frac{45}{72}=32\frac{61}{72}$
6. $\frac{1}{5}\text{x}\frac{1}{10}=\frac{1}{50}$
7. $\frac{10}{50}\div\frac{5}{50}=\frac{10\div5}{1}=2$

8. $\frac{1}{2} \times \frac{3}{5} = \frac{3}{10}$
9. $\frac{24}{32} = \frac{24}{32}$
10. $\frac{9}{90} < \frac{20}{90}$
11. $\frac{40}{88} > \frac{33}{88}$
12. $23 \div 4 = 5\frac{3}{4}$ gal
13. $5 \times 4 = 20$ qt
14. $30 \times 3 = 90$ ft
15. $5\frac{1}{2} + 3\frac{1}{2} = 8\frac{2}{2} = 9$ gallons total
    9-6=3 gallons left
    3x4=12 quarts left
16. $\frac{5}{16} \div \frac{1}{8} = \frac{40}{128} \div \frac{16}{128} =$
    $\frac{40 \div 16}{1} = \frac{40}{16} = \frac{5}{2} = 2\frac{1}{2}$ pieces
17. 2x2x2x2x3
18. $4 \times 4 = 16$
19. 1 in
20. $7 + 2 = 9$
    $9 \times 7 = 63$
    $63 + 2 = 65$
    $65 - 61 = 4$

9. $7\frac{3}{4} = \frac{31}{4}$
10. $10\frac{1}{8} = \frac{81}{8}$
11. $45\frac{1}{2} = \frac{91}{2}$
12. $24 \div 4 = 6$ gal
13. $20 \times 4 = 80$ qt
14. $15 \div 2 = 7\frac{1}{2}$ qt
15. $315 \div 35 = 9$ ft
16. $9 \div 3 = 3$ yd
17. no
18. 15 : $\underline{3}$, $\underline{5}$, $\underline{15}$
    45 : $\underline{3}$, $\underline{5}$, 9, $\underline{15}$, 45
    GCF = 15
19. $\frac{3}{4}$ of 24=18 want to go home
    $\frac{2}{3}$ of 18=12 have a ride
    18-12=6 will have to walk
20. $8 \times 6 = 48$
    $48 - 8 - 40$
    $40 \div 4 = 10$
    $10 \times 3 = 30$

Systematic Review 22F

1. $12\frac{14}{18} - 1\frac{9}{18} = 11\frac{5}{18}$
2. $4\frac{4}{3} - 2\frac{2}{3} = 2\frac{2}{3}$
3. $1\frac{3}{6} + 3\frac{4}{6} = 4\frac{7}{6} = 5\frac{1}{6}$
4. $4\frac{10}{15} + 3\frac{12}{15} = 7\frac{22}{15} = 8\frac{7}{15}$
5. $6\frac{24}{32} + 9\frac{28}{32} = 15\frac{52}{32} = 15\frac{13}{8} = 16\frac{5}{8}$
6. $\frac{12}{15} \div \frac{5}{15} = \frac{12 \div 5}{1} = \frac{12}{5} = 2\frac{2}{5}$
7. $\frac{3}{5} \times \frac{3}{7} = \frac{9}{35}$
8. $\frac{1}{4} \div \frac{3}{4} = \frac{1 \div 3}{1} = \frac{1}{3}$

Lesson Practice 23A

1. done
2. $\frac{3}{4} \times \frac{4}{3} = \frac{12}{12} = 1$
3. $\frac{1}{2} \times \frac{2}{1} = \frac{2}{2} = 1$
4. done
5. done
6. $\frac{8}{7}$
7. done
8. done
9. $\frac{3}{4} \times \frac{8}{5} = \frac{24}{20} = 1\frac{4}{20} = 1\frac{1}{5}$
10. $\frac{24}{32} \div \frac{20}{32} = \frac{24}{20} = 1\frac{4}{20} = 1\frac{1}{5}$

11. $\frac{9}{10} \times \frac{5}{1} = \frac{45}{10} = \frac{9}{2} = 4\frac{1}{2}$
12. $\frac{45}{50} \div \frac{10}{50} = \frac{45}{10} = \frac{9}{2} = 4\frac{1}{2}$
13. done
14. done
15. $\frac{11}{6} \div \frac{13}{4} = \frac{11}{6} \times \frac{4}{13} = \frac{44}{78} = \frac{22}{39}$
16. $\frac{15}{8} \div \frac{5}{8} = \frac{15}{8} \times \frac{8}{5} = \frac{120}{40} = 3$
17. $\frac{5}{3} \div \frac{5}{9} = \frac{5}{3} \times \frac{9}{5} = \frac{45}{15} = \frac{9}{3} = 3$ mi
18. $\frac{8}{3} \div \frac{1}{6} = \frac{8}{3} \times \frac{6}{1} = \frac{48}{3} = 16$ people

Lesson Practice 23B

1. $\frac{1}{3} \times \frac{3}{1} = \frac{3}{3} = 1$
2. $\frac{5}{8} \times \frac{8}{5} = \frac{40}{40} = 1$
3. $\frac{3}{7} \times \frac{7}{3} = \frac{21}{21} = 1$
4. $\frac{2}{1}$ or 2
5. $\frac{1}{9}$
6. $\frac{4}{3}$
7. $\frac{4}{1} \times \frac{2}{1} = \frac{8}{1} = 8$
8. $\frac{8}{2} \div \frac{1}{2} = \frac{8 \div 1}{1} = 8$
9. $\frac{7}{10} \times \frac{12}{7} = \frac{84}{70} = 1\frac{14}{70} = 1\frac{1}{5}$
10. $\frac{84}{120} \div \frac{70}{120} = \frac{84}{70} = 1\frac{14}{70} = 1\frac{1}{5}$
11. $\frac{3}{4} \times \frac{8}{1} = \frac{24}{4} = 6$
12. $\frac{24}{32} \div \frac{4}{32} = \frac{24 \div 4}{1} = 6$
13. $\frac{38}{5} \div \frac{3}{2} = \frac{38}{5} \times \frac{2}{3} = \frac{76}{15} = 5\frac{1}{15}$
14. $\frac{13}{4} \div \frac{74}{7} = \frac{13}{4} \times \frac{7}{74} = \frac{91}{296}$
15. $\frac{3}{2} \div \frac{1}{8} = \frac{3}{2} \times \frac{8}{1} = \frac{24}{2} = 12$
16. $\frac{13}{6} \div \frac{1}{6} = \frac{13}{6} \times \frac{6}{1} = \frac{78}{6} = 13$
17. $\frac{6}{5} \div \frac{9}{10} = \frac{6}{5} \times \frac{10}{9} =$
    $\frac{60}{45} = \frac{4}{3} = 1\frac{1}{3}$ ft
18. $\frac{5}{1} \div \frac{5}{4} = \frac{5}{1} \times \frac{4}{5} = \frac{20}{5} = 4$ times

Lesson Practice 23C

1. $\frac{1}{6} \times \frac{6}{1} = \frac{6}{6} = 1$
2. $\frac{2}{3} \times \frac{3}{2} = \frac{6}{6} = 1$
3. $\frac{4}{5} \times \frac{5}{4} = \frac{20}{20} = 1$
4. $\frac{6}{5}$
5. $\frac{1}{13}$
6. $\frac{4}{9}$
7. $\frac{5}{7} \times \frac{9}{2} = \frac{45}{14} = 3\frac{3}{14}$
8. $\frac{45}{63} \div \frac{14}{63} = \frac{45 \div 14}{1} = \frac{45}{14} = 3\frac{3}{14}$
9. $\frac{7}{8} \times \frac{3}{1} = \frac{21}{8} = 2\frac{5}{8}$
10. $\frac{21}{24} \div \frac{8}{24} = \frac{21 \div 8}{1} = \frac{21}{8} = 2\frac{5}{8}$
11. $\frac{2}{1} \times \frac{3}{2} = \frac{6}{2} = 3$
12. $\frac{6}{3} \div \frac{2}{3} = \frac{6 \div 2}{1} = 3$
13. $\frac{8}{3} \div \frac{11}{8} = \frac{8}{3} \times \frac{8}{11} = \frac{64}{33} = 1\frac{31}{33}$
14. $\frac{16}{3} \div \frac{8}{3} = \frac{16}{3} \times \frac{3}{8} = \frac{48}{24} = 2$
15. $\frac{14}{5} \div \frac{1}{10} = \frac{14}{5} \times \frac{10}{1} = \frac{140}{5} = 28$
16. $\frac{7}{4} \div \frac{5}{12} = \frac{7}{4} \times \frac{12}{5} = \frac{84}{20} = 4\frac{4}{20} = 4\frac{1}{5}$
17. $\frac{5}{2} \div \frac{1}{2} = \frac{5}{2} \times \frac{2}{1} = \frac{10}{2} = 5$ sections
18. $\frac{25}{4} \div \frac{5}{1} = \frac{25}{4} \times \frac{1}{5} = \frac{25}{20} = 1\frac{5}{20} = 1\frac{1}{4}$ yd

Systematic Review 23D

1. $\frac{1}{3} \times \frac{5}{1} = \frac{5}{3} = 1\frac{2}{3}$
2. $\frac{5}{15} \div \frac{3}{15} = \frac{5 \div 3}{1} = \frac{5}{3} = 1\frac{2}{3}$
3. $\frac{3}{4} \times \frac{8}{5} = \frac{24}{20} = \frac{6}{5} = 1\frac{1}{5}$
4. $\frac{24}{32} \div \frac{20}{32} = \frac{24 \div 20}{1} = \frac{24}{20} = \frac{6}{5} = 1\frac{1}{5}$
5. $\frac{9}{4} \div \frac{3}{5} = \frac{9}{4} \times \frac{5}{3} = \frac{45}{12} = \frac{15}{4} = 3\frac{3}{4}$
6. $\frac{11}{6} \div \frac{21}{10} = \frac{11}{6} \times \frac{10}{21} = \frac{110}{126} = \frac{55}{63}$
7. $9\frac{5}{30} - 4\frac{12}{30} = 8\frac{35}{30} - 4\frac{12}{30} = 4\frac{23}{30}$
8. $24\frac{4}{3} - 16\frac{2}{3} = 8\frac{2}{3}$
9. $7\frac{20}{32} + 2\frac{8}{32} = 9\frac{28}{32} = 9\frac{7}{8}$
10. done
11. done
12. $25 \div 16 = 1\frac{9}{16}$ lb
13. $10 \times 16 = 160$ oz
14. $\frac{35}{6} \div \frac{7}{6} = \frac{35}{6} \times \frac{6}{7} = \frac{210}{42} = 5$ pieces
15. $5\frac{4}{8} + 4\frac{6}{8} = 9\frac{10}{8}$

    $9\frac{10}{8} - 6\frac{3}{8} = 3\frac{7}{8}$ lb
16. $8 \times 4 \times 2 = 64$ cu. ft
17. $\frac{1}{2} + \frac{1}{2} = \frac{2}{2} = 1$

    $1 + \frac{3}{4} = 1\frac{3}{4}$ ft
18. $6 + 3 = 9$

    $9 - 4 = 5$

    $5 + 2 = 7$

    $7 \times 7 = 49$

Systematic Review 23E

1. $\frac{2}{3} \div \frac{5}{7} = \frac{2}{3} \times \frac{7}{5} = \frac{14}{15}$
2. $\frac{14}{21} \div \frac{15}{21} = \frac{14 \div 15}{1} = \frac{14}{15}$
3. $\frac{9}{10} \div \frac{1}{5} = \frac{9}{10} \times \frac{5}{1} = \frac{45}{10} = \frac{9}{2} = 4\frac{1}{2}$
4. $\frac{45}{50} \div \frac{10}{50} = \frac{45 \div 10}{1} = \frac{45}{10} = \frac{9}{2} = 4\frac{1}{2}$
5. $\frac{23}{5} \div \frac{17}{7} = \frac{23}{5} \times \frac{7}{17} = \frac{161}{85} = 1\frac{76}{85}$
6. $\frac{9}{5} \div \frac{4}{3} = \frac{9}{5} \times \frac{3}{4} = \frac{27}{20} = 1\frac{7}{20}$
7. $4\frac{6}{8} - 1\frac{4}{8} = 3\frac{2}{8} = 3\frac{1}{4}$
8. $13\frac{6}{21} - 11\frac{7}{21} = 12\frac{27}{21} - 11\frac{7}{21} = 1\frac{20}{21}$
9. $3\frac{1}{3} + 2\frac{2}{3} = 5\frac{3}{3} = 6$
10. $\frac{15}{20} + \frac{12}{20} = \frac{27}{20}$

    $\frac{27}{20} + \frac{6}{10} = \frac{270}{200} + \frac{120}{200} = \frac{390}{200} =$

    $\frac{39}{20} = 1\frac{19}{20}$
11. $\frac{7}{21} + \frac{15}{21} = \frac{22}{21}$

    $\frac{22}{21} + \frac{2}{5} = \frac{110}{105} + \frac{42}{105} = \frac{152}{105} = 1\frac{47}{105}$
12. $\frac{3}{6} + \frac{4}{6} = \frac{7}{6}$

    $\frac{7}{6} + \frac{4}{7} = \frac{49}{42} + \frac{24}{42} = \frac{73}{42} = 1\frac{31}{42}$
13. $64 \div 16 = 4$ lb
14. $6 \times 16 = 96$ oz
15. $13 \div 4 = 3\frac{1}{4}$ gal
16. $\frac{1}{2} \times 20 \times 5 = 50$ sq. ft
17. $10\frac{1}{2} + 7\frac{1}{2} = 17\frac{2}{2} = 18$ ft; $18 \div 3 = 6$ yd
18. $\frac{93}{10} \div \frac{12}{5} = \frac{93}{10} \times \frac{5}{12} = \frac{465}{120} = 3\frac{105}{120} =$

    $3\frac{7}{8}$ ft
19. $7 \times 7 \times 7 = 343$ cu. ft
20. $11 - 4 = 7$

    $7 + 3 = 10$

    $10 + 12 = 22$

    $22 \times 2 = 44$

Systematic Review 23F

1. $\frac{5}{9} \div \frac{2}{3} = \frac{5}{9} x \frac{3}{2} = \frac{15}{18} = \frac{5}{6}$
2. $\frac{15}{27} \div \frac{18}{27} = \frac{15 \div 18}{1} = \frac{15}{18} = \frac{5}{6}$
3. $\frac{1}{2} \div \frac{1}{3} = \frac{1}{2} x \frac{3}{1} = \frac{3}{2} = 1\frac{1}{2}$
4. $\frac{3}{6} \div \frac{2}{6} = \frac{3 \div 2}{1} = \frac{3}{2} = 1\frac{1}{2}$
5. $\frac{11}{4} \div \frac{5}{8} = \frac{11}{4} x \frac{8}{5} = \frac{88}{20} = 4\frac{8}{20} = 4\frac{2}{5}$
6. $\frac{12}{5} \div \frac{7}{6} = \frac{12}{5} x \frac{6}{7} = \frac{72}{35} = 2\frac{2}{35}$
7. $11\frac{8}{7} - 6\frac{4}{7} = 5\frac{4}{7}$
8. $3\frac{2}{16} - 2\frac{8}{16} = 2\frac{18}{16} - 2\frac{8}{16} = \frac{10}{16} = \frac{5}{8}$
9. $5\frac{5}{50} + 2\frac{20}{50} = 7\frac{25}{50} = 7\frac{1}{2}$
10. $\frac{3}{4} = \frac{6}{8} = \frac{9}{12}$
11. $\frac{1}{8} = \frac{2}{16} = \frac{3}{24}$
12. $\frac{7}{10} = \frac{14}{20} = \frac{21}{30}$
13. $21 \div 16 = \frac{21}{16} = 1\frac{5}{16}$ lb
14. $14 \div 2 = 7$ qt
15. $25x3 = 75$ ft
16. $\frac{15}{4} \div \frac{5}{1} = \frac{15}{4} x \frac{1}{5} = \frac{15}{20} = \frac{3}{4}$ lb
17. $14x11x9 = 1{,}386$ cu. ft
18. $\frac{1}{10}$
19. $\frac{3}{4}$ of 2 = 15 gal used
    20-15=5 gal left
    5x4=20 qt left
20. $15 \div 5 = 3$
    $3 + 5 = 8$
    $8x6 = 48$
    $48 + 2 = 50$

Lesson Practice 24A

1. done
2. done
3. $\frac{10}{9}$
4. $\frac{1}{7}$
5. done
6. done
7. $\frac{1}{8} \cdot 8A = 40 \cdot \frac{1}{8}$
   $A = 5$
8. $8(5) = 40$
   $40 = 40$
9. $\frac{1}{3} \cdot 3D = 27 \cdot \frac{1}{3}$
   $D = 9$
10. $3(9) = 27$
    $27 = 27$
11. $\frac{1}{11} \cdot 11Y = 121 \cdot \frac{1}{11}$
    $Y = 11$
12. $11(11) = 121$
    $121 = 121$
13. $\frac{1}{3} \cdot 3R = 12 \cdot \frac{1}{3}$
    R = 4 rabbits
14. $\frac{1}{4} \cdot 4D = 48 \cdot \frac{1}{4}$
    D = $12

Lesson Practice 24B

1. $\frac{3}{2}$
2. $\frac{1}{10}$
3. $\frac{8}{1} = 8$
4. $\frac{1}{2}$
5. $\frac{1}{8} \cdot 8Z = 72 \cdot \frac{1}{8}$
   $Z = 9$
6. $8(9) = 72$
   $72 = 72$

7. $\frac{1}{10} \cdot 10C = 100 \cdot \frac{1}{10}$
   $C = 10$
8. $10(10) = 100$
   $100 = 100$
9. $\frac{1}{5} \cdot 5F = 25 \cdot \frac{1}{5}$
   $F = 5$
10. $5(5) = 25$
    $25 = 25$
11. $\frac{1}{2} \cdot 2A = 6 \cdot \frac{1}{2}$
    $A = 3$
12. $2(3) = 6$
    $6 = 6$
13. $\frac{1}{2} \cdot 2A = 12 \cdot \frac{1}{2}$
    $A = 6$ years old
14. $\frac{1}{5} \cdot 5B = 15 \cdot \frac{1}{5}$
    $B = 3$ books

Lesson Practice 24C

1. $\frac{9}{7}$
2. $\frac{1}{8}$
3. $\frac{2}{3}$
4. $\frac{1}{32}$
5. $\frac{1}{4} \cdot 4B = 8 \cdot \frac{1}{4}$
   $B = 2$
6. $4(2) = 8$
   $8 = 8$
7. $\frac{1}{5} \cdot 5E = 20 \frac{1}{5}$
   $E = 4$
8. $5(4) = 20$
   $20 = 20$
9. $\frac{1}{2} \cdot 2H = 24 \cdot \frac{1}{2}$
   $H = 12$
10. $2(12) = 24$
    $24 = 24$
11. $\frac{1}{3} \cdot 3C = 63 \cdot \frac{1}{3}$
    $C = 21$
12. $3(21) = 63$
    $63 = 63$
13. $5A = 50$
    $\frac{1}{5} \cdot 5A = 5 \cdot \frac{1}{5}$
    $A = 10$ years old
    Student may have used a different letter.
14. $9X = 81$
    $\frac{1}{9} \cdot 9X = 8 \cdot \frac{1}{9}$
    $X = 9$ ft

Systematic Review 24D

1. $\frac{1}{7} \cdot 7F = 49 \cdot \frac{1}{7}$
   $F = 7$
2. $7(7) = 49$
   $49 = 49$
3. $\frac{1}{12} \cdot 12R = 36 \cdot \frac{1}{12}$
   $R = 3$
4. $12(3) = 36$
   $36 = 36$
5. $\frac{1}{2} \div \frac{1}{4} = \frac{1}{2} \times \frac{4}{1} = \frac{4}{2} = 2$
6. $\frac{4}{8} \div \frac{2}{8} = \frac{4 \div 2}{1} = 2$
7. $\frac{2}{5} \div \frac{11}{6} = \frac{2}{5} \times \frac{6}{11} = \frac{12}{55}$
8. $\frac{11}{2} \div \frac{22}{9} = \frac{11}{2} \times \frac{9}{22} = \frac{99}{44} = \frac{9}{4} = 2\frac{1}{4}$
9. $8\frac{10}{35} - 3\frac{28}{35} = 7\frac{45}{35} - 3\frac{28}{35} = 4\frac{17}{35}$
10. $16\frac{5}{8} - 10\frac{3}{8} = 6\frac{2}{8} = 6\frac{1}{4}$
11. $5\frac{14}{20} + 4\frac{10}{20} = 9\frac{24}{20} = 10\frac{4}{20} = 10\frac{1}{5}$
12. done
13. done

14. $60 \div 12 = 5$ ft
15. $8 \times 12 = 96$ in
16. $4X = 16$
$\frac{1}{4} \cdot 4X = 16 \cdot \frac{1}{4}$
$X = 4$ gifts
17. $5 \times 16 = 80$ oz
18. $9 \times 9 = 81$
$81 - 1 = 80$
$80 \div 2 = 40$
$40 \div 5 = 8$

Systematic Review 24E

1. $\frac{1}{6} \cdot 6G = 42 \cdot \frac{1}{6}$
$G = 7$
2. $6(7) = 42$
$42 = 42$
3. $\frac{1}{9} \cdot 9S = 54 \cdot \frac{1}{9}$
$S = 6$
4. $9(6) = 54$
$54 = 54$
5. $\frac{6}{8} \div \frac{1}{2} = \frac{6}{8} \times \frac{2}{1} = \frac{12}{8} = \frac{3}{2} = 1\frac{1}{2}$
6. $\frac{12}{16} \div \frac{8}{16} = \frac{12 \div 8}{1} = \frac{12}{8} = \frac{3}{2} = 1\frac{1}{2}$
7. $\frac{34}{5} \div \frac{5}{3} = \frac{34}{5} \times \frac{3}{5} = \frac{102}{25} = 4\frac{2}{25}$
8. $\frac{25}{3} \div \frac{45}{8} = \frac{25}{3} \times \frac{8}{45} = \frac{200}{135} =$
$1\frac{65}{135} = 1\frac{13}{27}$
9. $4\frac{1}{4} + 2\frac{3}{4} = 6\frac{4}{4} = 7$
10. $2\frac{5}{50} + 6\frac{10}{50} = 8\frac{15}{50} = 8\frac{3}{10}$
11. $13\frac{9}{18} - 5\frac{14}{18} = 12\frac{27}{18} - 5\frac{14}{18} = 7\frac{13}{18}$
12. $\frac{5}{6} \times \frac{2}{3} = \frac{10}{18} = \frac{5}{9}$
13. $\frac{4}{5} \times \frac{2}{7} = \frac{8}{35}$
14. $\frac{1}{9} \times \frac{3}{8} = \frac{3}{72} = \frac{1}{24}$
15. $3S = 45$; $S = 15$ years
16. $A = \frac{5}{8} \times \frac{5}{8} = \frac{25}{64}$ sq in
$P = \frac{5}{8} + \frac{5}{8} + \frac{5}{8} + \frac{5}{8} = \frac{20}{8} = \frac{5}{2} = 2\frac{1}{2}$ in
17. $32 \div 3 = 10\frac{2}{3}$ yd
18. $32 \times 12 = 384$ in
19. $15 \times 15 = 225$
20. $30 - 10 = 20$
$20 - 2 = 18$
$18 \div 6 = 3$
$3 \times 7 = 21$

Systematic Review 24F

1. $\frac{1}{8} \cdot 8H = 72 \cdot \frac{1}{8}$
$H = 9$
2. $8(9) = 72$
$72 = 72$
3. $\frac{1}{11} \cdot 11T = 55 \cdot \frac{1}{11}$
$T = 5$
4. $11(5) = 55$
$55 = 55$
5. $\frac{1}{4} \div \frac{1}{3} = \frac{1}{4} \times \frac{3}{1} = \frac{3}{4}$
6. $\frac{3}{12} \div \frac{4}{12} = \frac{3 \div 4}{1} = \frac{3}{4}$
7. $\frac{7}{2} \div \frac{13}{3} = \frac{7}{2} \times \frac{3}{13} = \frac{21}{26}$
8. $\frac{23}{5} \div \frac{5}{3} = \frac{23}{5} \times \frac{3}{5} = \frac{69}{25} = 2\frac{19}{25}$
9. $8\frac{3}{4} - 8\frac{1}{4} = \frac{2}{4} = \frac{1}{2}$
10. $3\frac{14}{16} + 2\frac{8}{16} = 5\frac{22}{16} = 6\frac{6}{16} = 6\frac{3}{8}$
11. $7\frac{16}{40} - 3\frac{25}{40} = 6\frac{56}{40} - 3\frac{25}{40} = 3\frac{31}{40}$
12. $\frac{3}{4} \times \frac{1}{8} = \frac{3}{32}$
13. $\frac{1}{3} \times \frac{3}{7} = \frac{3}{21} = \frac{1}{7}$
14. $\frac{2}{5} \times \frac{5}{6} = \frac{10}{30} = \frac{1}{3}$
15. $6D = 18$; $D = 3$ dimes
16. $5 + 10 + 5 + 10 = 30$ ft
$30 \div 3 = 10$ yd

17. $5\frac{4}{8} - 4\frac{6}{8} = 4\frac{12}{8} - 4\frac{6}{8} = \frac{6}{8} = \frac{3}{4}$ gal
18. $\frac{3}{4}$ of $12 = 9$ in
19. $\frac{3}{4} \times 4 = \frac{12}{4} = 3$ qt
20. $20 \div 4 = 5$
    $5 + 4 = 9$
    $9 \times 2 = 18$
    $18 \div 6 = 3$

Lesson Practice 25A

1. done
2. done
3. $\frac{1}{_2\cancel{6}} \times \frac{^1\cancel{3}}{8} = \frac{1}{16}$
4. $\frac{1}{_1\cancel{2}} \times \frac{^2\cancel{4}}{5} = \frac{2}{5}$
5. $\frac{1}{_1\cancel{4}} \times \frac{^1\cancel{7}}{11} \times \frac{^1\cancel{4}}{\cancel{7}} = \frac{1}{11}$
6. $\frac{^1\cancel{4}}{_1\cancel{5}} \times \frac{1}{2} \times \frac{^1\cancel{5}}{_2\cancel{8}} = \frac{1}{4}$
7. $\frac{1}{_1\cancel{5}} \times \frac{^1\cancel{5}}{6} = \frac{1}{6}$
8. $\frac{1}{_1\cancel{2}} \times \frac{^2\cancel{6}}{7} \times \frac{^1\cancel{2}}{\cancel{3}_1} = \frac{2}{7}$
9. $\frac{1}{_2\cancel{4}} \times \frac{^1\cancel{3}}{5} \times \frac{^1\cancel{2}}{_1\cancel{3}} = \frac{1}{10}$
10. done
11. $\frac{11}{_1\cancel{5}} \times \frac{^2\cancel{10}}{7} \times \frac{8}{3} = \frac{176}{21} = 8\frac{8}{21}$
12. $\frac{^3\cancel{12}}{_1\cancel{7}} \times \frac{^1\cancel{7}}{_1\cancel{4}} = \frac{3}{1} = 3$
13. $\frac{^1\cancel{5}}{6} \times \frac{^1\cancel{7}}{_1\cancel{5}} \times \frac{1}{_1\cancel{7}} = \frac{1}{6}$
14. $\frac{1}{2} \times \frac{1}{2} \times \frac{1}{2} = \frac{1}{8}$ of a pie
15. $\frac{^1\cancel{2}}{_1\cancel{3}} \times \frac{1}{_1\cancel{2}} \times \frac{^1\cancel{3}}{4} = \frac{1}{4}$ of the snacks

Lesson Practice 25B

1. $\frac{4}{_3\cancel{6}} \times \frac{^1\cancel{2}}{5} = \frac{4}{15}$
2. $\frac{1}{_3\cancel{6}} \times \frac{^1\cancel{3}}{11} \times \frac{^1\cancel{2}}{_1\cancel{3}} = \frac{1}{33}$
3. $\frac{^1\cancel{3}}{_4\cancel{8}} \times \frac{^1\cancel{2}}{5} \times \frac{3}{\cancel{3}} = \frac{3}{20}$
4. $\frac{4}{5} \times \frac{^1\cancel{2}}{_3\cancel{6}} = \frac{4}{15}$
5. $\frac{^1\cancel{2}}{5} \times \frac{^1\cancel{7}}{_1\cancel{4}\,8} \times \frac{^1\cancel{4}}{_1\cancel{7}} = \frac{1}{5}$
6. $\frac{1}{_1\cancel{3}} \times \frac{^1\cancel{4}}{9} \times \frac{^1\cancel{3}}{_2\cancel{8}} = \frac{1}{18}$
7. $\frac{1}{_1\cancel{2}} \times \frac{^1\cancel{2}\,\cancel{4}}{_3\cancel{6}} = \frac{1}{3}$
8. $\frac{1}{_1\cancel{3}} \times \frac{^2\cancel{6}}{7} \times \frac{1}{5} = \frac{2}{35}$
9. $\frac{^1\cancel{5}}{8} \times \frac{1}{_1\cancel{2}} \times \frac{^1\cancel{2}}{_1\cancel{5}} = \frac{1}{8}$
10. $\frac{^1\cancel{5}\,\cancel{25}}{_3\cancel{6}} \times \frac{^2\cancel{4}\,\cancel{8}}{_1\cancel{5}} \times \frac{1}{\cancel{10}\,\cancel{2}_1} = \frac{2}{3}$
11. $\frac{1}{_1\cancel{2}} \times \frac{^1\cancel{2}\,\cancel{4}}{_1\cancel{11}} \times \frac{^1\cancel{11}}{_3\cancel{6}} = \frac{1}{3}$
12. $\frac{1}{_1\cancel{6}} \times \frac{^1\cancel{7}}{_2\cancel{4}} \times \frac{^1\cancel{2}\,\cancel{12}}{_1\cancel{7}} = \frac{1}{2}$
13. $\frac{^3\cancel{9}}{2} \times \frac{^1\cancel{5}}{_1\cancel{3}} \times \frac{11}{_1\cancel{5}} = \frac{33}{2} = 16\frac{1}{2}$

14. $\frac{{}^1\cancel{5}}{2} \times \frac{{}^1\cancel{3}}{{}_1\cancel{5}} \times \frac{1}{{}_1\cancel{3}} = \frac{1}{2}$ of a bushel

15. $\frac{{}^1\cancel{7}}{{}_2\cancel{10}} \times \frac{{}^1\cancel{5}}{6} \times \frac{1}{{}_1\cancel{7}} = \frac{1}{12}$ of the profit

$144 \div 12 = 12$

12x1 = $12 for the son

13. $\frac{{}^1\cancel{2}\,\cancel{6}}{{}_1\cancel{5}} \times \frac{{}^1\cancel{2}\,\cancel{10}}{{}_1\cancel{3}} \times \frac{7}{{}_1\cancel{2}\,\cancel{4}} = \frac{7}{1} = 7$

14. $\frac{1}{{}_1\cancel{5}} \times \frac{{}^1\cancel{3}}{8} \times \frac{{}^1\cancel{5}}{{}_2\cancel{6}} = \frac{1}{16}$ of total income

15. $\frac{{}^1\cancel{3}\,9}{2} \times \frac{1}{{}_1\cancel{3}} \times \frac{1}{{}_1\cancel{3}} = \frac{1}{2}$ of a pie

Lesson Practice 25C

1. $\frac{{}^1\cancel{4}}{5} \times \frac{1}{{}_2\cancel{8}} = \frac{1}{10}$
2. $\frac{{}^1\cancel{2}}{{}_1\cancel{3}} \times \frac{{}^1\cancel{5}}{{}_3\cancel{6}} \times \frac{{}^1\cancel{3}}{{}_1\cancel{5}} = \frac{1}{3}$
3. $\frac{{}^1\cancel{3}}{4} \times \frac{1}{{}_1\cancel{5}} \times \frac{{}^1\cancel{5}}{{}_3\cancel{9}} = \frac{1}{12}$
4. $\frac{{}^1\cancel{7}}{{}_5\cancel{10}} \times \frac{{}^1\cancel{2}}{{}_3\cancel{21}} = \frac{1}{15}$
5. $\frac{{}^1\cancel{5}}{{}_3\cancel{9}} \times \frac{5}{6} \times \frac{{}^1\cancel{3}}{{}_1\cancel{5}} = \frac{5}{18}$
6. $\frac{7}{{}_1\cancel{4}\,\cancel{8}} \times \frac{{}^1\cancel{2}}{3} \times \frac{{}^1\cancel{4}}{5} = \frac{7}{15}$
7. $\frac{5}{{}_2\cancel{6}} \times \frac{{}^1\cancel{3}}{4} = \frac{5}{8}$
8. $\frac{3}{5} \times \frac{{}^1\cancel{7}}{{}_5\cancel{10}} \times \frac{{}^2\cancel{4}}{{}_1\cancel{7}} = \frac{6}{25}$
9. $\frac{{}^1\cancel{5}}{{}_1\cancel{4}\,\cancel{8}} \times \frac{{}^1\cancel{2}}{3} \times \frac{{}^1\cancel{4}}{{}_1\cancel{5}} = \frac{1}{3}$
10. $\frac{{}^1\cancel{3}}{{}_1\cancel{8}} \times \frac{{}^1\cancel{8}}{{}_1\cancel{3}} \times \frac{2}{3} = \frac{2}{3}$
11. $\frac{{}^1\cancel{5}}{{}_2\cancel{6}} \times \frac{{}^1\cancel{2}}{{}_1\cancel{5}} \times \frac{{}^1\cancel{3}}{{}_1\cancel{2}} = \frac{1}{2}$
12. $\frac{{}^1\cancel{5}}{{}_1\cancel{9}} \times \frac{{}^2\cancel{14}}{{}_1\cancel{5}} \times \frac{{}^2\cancel{18}}{{}_1\cancel{7}} = \frac{4}{1} = 4$

Systematic Review 25D

1. $\frac{{}^{\cancel{2}}\cancel{4}}{\cancel{7}} \times \frac{\cancel{7}}{\cancel{2}} \times \frac{5}{{}_3\cancel{6}} = \frac{5}{3} = 1\frac{2}{3}$

   (When the result of reducing is a 1, it is not necessary to write it each time.)
2. $\frac{\cancel{2}}{{}_3\cancel{9}} \times \frac{\cancel{3}}{\cancel{5}} \times \frac{\cancel{5}}{\cancel{2}} = \frac{1}{3}$
3. $\frac{1}{9} \cdot 9G = 36 \cdot \frac{1}{9}$

   $G = 4$
4. $9(4) = 36$

   $36 = 36$
5. $\frac{1}{12} \cdot 12V = 144 \cdot \frac{1}{12}$

   $V = 12$
6. $12(12) = 144$

   $144 = 144$
7. $\frac{27}{5} \div \frac{17}{5} = \frac{27 \div 17}{1} = \frac{27}{17} = 1\frac{10}{17}$
8. $\frac{82}{7} \div \frac{8}{7} = \frac{{}^{41}\cancel{82}}{\cancel{7}} \times \frac{\cancel{7}}{{}_4\cancel{8}} = \frac{41}{4} = 10\frac{1}{4}$
9. $5\frac{1}{4} - 2\frac{3}{4} = 4\frac{5}{4} - 2\frac{3}{4} = 2\frac{2}{4} = 2\frac{1}{2}$
10. $7\frac{20}{50} - 3\frac{35}{50} = 6\frac{70}{50} - 3\frac{35}{50}$

    $= 3\frac{35}{50} = 3\frac{7}{10}$
11. $9\frac{4}{12} + 6\frac{3}{12} = 15\frac{7}{12}$
12. done
13. done
14. $3 \times 2{,}000 = 6{,}000$ lb

15. $\frac{\cancel{5}}{8} \times \frac{1}{2} \times \frac{1}{\cancel{5}} = \frac{1}{16}$ of the pizza

16. $6X = 300$

    $\frac{1}{6} \cdot 6X = 300 \cdot \frac{1}{6}$

    $X = 50$ years old

17. $\frac{3}{4} \times 2{,}000 = \frac{6{,}000}{4} = 1{,}500$ lb

    or $\frac{3}{4}$ of 2,000:

    $2{,}000 \div 4 = 500$; $500 \times 3 = 1{,}500$ lb

18. $40 - 5 = 35$

    $35 \div 7 = 5$

    $5 + 3 = 8$

    $8 \times 7 = 56$

Systematic Review 25E

1. $\frac{1}{\cancel{3}} \times \frac{\overset{5}{\cancel{15}}}{\underset{2}{\cancel{8}}} \times \frac{\cancel{4}}{\cancel{5}} = \frac{1}{2}$

2. $\frac{17}{\cancel{6}} \times \frac{\cancel{6}}{\cancel{5}} \times \frac{\overset{5}{\cancel{25}}}{8} = \frac{85}{8} = 10\frac{5}{8}$

3. $\frac{1}{7} \cdot 7H = 140 \cdot \frac{1}{7}$

   $H = 20$

4. $7(20) = 140$

   $140 = 140$

5. $\frac{1}{16} \cdot 16W = 48 \cdot \frac{1}{16}$

   $W = 3$

6. $16(3) = 48$

   $48 = 48$

7. $\frac{13}{4} \div \frac{21}{5} = \frac{13}{4} \times \frac{5}{21} = \frac{65}{84}$

8. $\frac{4}{11} \div \frac{2}{11} = \frac{4 \div 2}{1} = 2$

9. $4\frac{2}{12} - 1\frac{6}{12} = 3\frac{14}{12} - 1\frac{6}{12} = 2\frac{8}{12} = 2\frac{2}{3}$

10. $8\frac{4}{8} - 2\frac{6}{8} = 7\frac{12}{8} - 2\frac{6}{8} = 5\frac{6}{8} = 5\frac{3}{4}$

11. $7\frac{12}{15} + 2\frac{5}{15} = 9\frac{17}{15} = 10\frac{2}{15}$

12. $\frac{2}{3} = \frac{4}{6} = \frac{6}{9} = \frac{8}{12}$

13. $\frac{1}{10} = \frac{2}{20} = \frac{3}{30} = \frac{4}{40}$

14. $7 \times 2{,}000 = 14{,}000$ lb

15. $9{,}000 \div 2{,}000 = 4\frac{1000}{2000} = 4\frac{1}{2}$ tons

16. $\frac{1}{2} \times \frac{2000}{1} = \frac{2000}{2} = 1{,}000$ lb

17. $\frac{\overset{2}{\cancel{10}}}{\cancel{3}} \times \frac{1}{3} \times \frac{\cancel{3}}{\cancel{5}} = \frac{2}{3}$ acre

18. $\frac{5}{2} \times \frac{4}{1} = \frac{20}{2} = 10$ qt

19. $\frac{35}{4} \div \frac{5}{1} = \frac{\overset{7}{\cancel{35}}}{4} \times \frac{1}{\cancel{5}} = \frac{7}{4} = 1\frac{3}{4}$ yd

20. $60 + 3 = 63$

    $63 \div 9 = 7$

    $7 + 4 = 11$

    $11 \times 3 = 33$

Systematic Review 25F

1. $\frac{3}{\cancel{5}} \times \frac{\cancel{5}}{4} \times \frac{1}{4} = \frac{3}{16}$

2. $\frac{\overset{5}{\cancel{15}}}{\cancel{8}} \times \frac{5}{6} \times \frac{\cancel{8}}{\cancel{3}} = \frac{25}{6} = 4\frac{1}{6}$

3. $\frac{1}{6} \cdot 6K = 54 \cdot \frac{1}{6}$

   $K = 9$

4. $6(9) = 54$

   $54 = 54$

5. $\frac{1}{25} \cdot 25X = 450 \cdot \frac{1}{25}$

   $X = 18$

6. $25(18) = 450$

7. $\frac{16}{9} \div \frac{5}{3} = \frac{16}{\underset{3}{\cancel{9}}} \times \frac{\cancel{3}}{5} = \frac{16}{15} = 1\frac{1}{15}$

8. $\frac{7}{8} \div \frac{1}{8} = \frac{7 \div 1}{1} = 7$

9. $10\frac{4}{12} - 3\frac{9}{12} = 9\frac{16}{12} - 3\frac{9}{12} = 6\frac{7}{12}$

10. $6\frac{10}{15} - 2\frac{12}{15} = 5\frac{25}{15} - 2\frac{12}{15} = 3\frac{13}{15}$

11. $9\frac{4}{6} + 7\frac{5}{6} = 16\frac{9}{6} = 17\frac{3}{6} = 17\frac{1}{2}$

12. $\frac{4}{5}=\frac{8}{10}=\frac{12}{15}=\frac{16}{20}$
13. $\frac{3}{7}=\frac{6}{14}=\frac{9}{21}=\frac{12}{28}$
14. $4\text{x}2{,}000=8{,}000$ lb
15. $7\div 2=3\frac{1}{2}$ qt
16. $\frac{3}{\cancel{4}}\text{x}\frac{\overset{4}{\cancel{16}}}{1}=\frac{12}{1}=12$ oz
17. $\frac{11}{\cancel{2}}\text{x}\frac{\overset{6}{\cancel{12}}}{1}=\frac{66}{1}=66$ in
18. $\frac{11}{2}\div\frac{3}{1}=\frac{11}{2}\text{x}\frac{1}{3}=\frac{11}{6}=1\frac{5}{6}$ yd
19. $\frac{1}{3}\text{x}\frac{1}{2}\text{x}\frac{1}{4}=\frac{1}{24}$ of original price
20. $8\text{x}6=48$
    $48+2=50$
    $50\div 5=10$
    $10\text{x}10=100$

Lesson Practice 26A

1. done
2. done
3. $4B+8=28$
   $-8=-8$
   $4B=20$
   $\frac{1}{4}\cdot 4B=20\cdot\frac{1}{4}$
   $B=5$
4. $4(5)+8=28$
   $20+8=28$
   $28=28$
5. $2Y-11=7$
   $+11\quad +11$
   $2Y=18$
   $\frac{1}{2}\cdot 2Y=18\cdot\frac{1}{2}$
   $Y=9$
6. $2(9)-11=7$
   $18-11=7$
   $7=7$
7. $5C+8=18$
   $-8\quad -8$
   $5C=10$
   $\frac{1}{5}\cdot 5C=10\cdot\frac{1}{5}$
   $C=2$
8. $5(2)+8=18$
   $10+8=18$
   $18=18$
9. $5Z-2=38$
   $+2\quad +2$
   $5Z=40$
   $\frac{1}{5}\cdot 5Z=40\cdot\frac{1}{5}$
   $Z=8$
10. $5(8)-2=38$
    $40-2=38$
    $38=38$
11. $3D+9=21$
    $-9\quad -9$
    $3D=12$
    $\frac{1}{3}\cdot 3D=12\cdot\frac{1}{3}$
    $D=4$
12. $3(4)+9=21$
    $12+9=21$
    $21=21$
13. $7A-7=42$
    $+7\quad +7$
    $7A=49$
    $\frac{1}{7}\cdot 7A=49\cdot\frac{1}{7}$
    $A=7$
14. $7(7)-7=42$
    $49-7=42$
    $42=42$

Lesson Practice 26B

1. $5B+3=18$
   $-3=-3$
   $5B=15$

$\frac{1}{5}\cdot 5B = 15\cdot\frac{1}{5}$
$B = 3$

2. $5(3)+3=18$
$15+3=18$
$18=18$

3. $12T-1=35$
$\underline{+1 \;=+1}$
$12T \;=36$
$\frac{1}{12}\cdot 12T = 36\cdot\frac{1}{12}$
$T=3$

4. $12(3)-1=35$
$36-1=35$
$35=35$

5. $4C+7= \;31$
$\underline{-7 \;\; -7}$
$4C \;= 24$
$\frac{1}{4}\cdot 4C = 24\cdot\frac{1}{4}$
$C=6$

6. $4(6)+7=31$
$24+7=31$
$31=31$

7. $6S-13= \;11$
$\underline{+13 \;\; +13}$
$6S \;=24$
$\frac{1}{6}\cdot 6S = 24\cdot\frac{1}{6}$
$S=4$

8. $6(4)-13=11$
$24-13=11$
$11=11$

9. $10D+11= \;91$
$\underline{-11 \;\; -11}$
$10D \;= 80$
$\frac{1}{10}\cdot 10D = 80\cdot\frac{1}{10}$
$D=8$

10. $10(8)+11=91$
$80+11=91$
$91=91$

11. $7W-13=22$
$\underline{+13 \;\; +13}$
$7W \;= 35$
$\frac{1}{7}\cdot 7W = 35\cdot\frac{1}{7}$
$W=5$

12. $7(5)-13=22$
$35-13=22$
$22=22$

13. $3E+17=20$
$\underline{-17 \;\; -17}$
$3E \;= \;3$
$\frac{1}{3}\cdot 3E = 3\cdot\frac{1}{3}$
$E=1$

14. $3(1)+17=20$
$3+17=20$
$20=20$

Lesson Practice 26C

1. $5F+13=38$
$\underline{-13=-13}$
$5F \;=25$
$\frac{1}{5}\cdot 5F = 25\cdot\frac{1}{5}$
$F=5$

2. $5(5)+13=38$
$25+13=38$
$38=38$

3. $8V-14= \;26$
$\underline{+14=+14}$
$8V \;= \;40$
$\frac{1}{8}\cdot 8V = 40\cdot\frac{1}{8}$
$V=5$

4. $8(5)-14=26$
$40-14=26$
$26=26$

5. $7G+21= \;56$
$\underline{-21 \;\; -21}$
$7G \;= \;35$
$\frac{1}{7}\cdot 7G = 35\cdot\frac{1}{7}$
$G=5$

6. $7(5)+21=56$
   $35+21=56$
   $56=56$
7. $5W-12=53$
   $+12\ \ +12$
   $5W=65$
   $\frac{1}{5}\cdot 5W=65\cdot\frac{1}{5}$
   $W=13$
8. $5(13)-12=53$
   $65-12=53$
   $53=53$
9. $6H+7=25$
   $-7\ \ -7$
   $6H=18$
   $\frac{1}{6}\cdot 6H=18\cdot\frac{1}{6}$
   $H=3$
10. $6(3)+7=25$
    $18+7=25$
    $25=25$
11. $11X-16=72$
    $+16\ \ +16$
    $11X=88$
    $\frac{1}{11}\cdot 11X=88\cdot\frac{1}{11}$
    $X=8$
12. $11(8)-16=72$
    $88-16=72$
    $72=72$
13. $2J+17=37$
    $-17\ \ -17$
    $2J=20$
    $\frac{1}{2}\cdot 2J=20\cdot\frac{1}{2}$
    $J=10$
14. $2(10)+17=37$
    $20+17=37$
    $37=37$

Systematic Review 26D

1. $6Y-12=12$
   $6Y=24$
   $\frac{1}{6}\cdot 6=24\cdot\frac{1}{6}$
   $Y=4$
2. $6(4)-12=12$
   $24-12=12$
   $12=12$
3. $4K+6=46$
   $4K=40$
   $\frac{1}{4}\cdot 4K=40\cdot\frac{1}{4}$
   $K=10$
4. $4(10)+6=46$
   $40+6=46$
   $46=46$
5. $\frac{\overset{5}{\cancel{25}}}{3}\times\frac{\overset{2}{\cancel{34}}}{\cancel{5}}\times\frac{1}{\cancel{17}}=\frac{10}{3}=3\frac{1}{3}$
6. $\frac{\cancel{3}}{\cancel{11}}\times\frac{\cancel{11}}{\cancel{5}}\times\frac{\overset{2}{\cancel{10}}}{\cancel{3}}=\frac{2}{1}=2$
7. $\frac{5}{16}\div\frac{2}{8}=\frac{5}{\underset{2}{\cancel{16}}}\times\frac{\cancel{8}}{2}=\frac{5}{4}=1\frac{1}{4}$
8. $\frac{23}{5}\div\frac{13}{6}=\frac{23}{5}\times\frac{6}{13}=\frac{138}{65}=2\frac{8}{65}$
9. $\frac{6}{12}+\frac{10}{12}=\frac{16}{12}=\frac{4}{3}$
   $\frac{4}{3}+\frac{5}{9}=\frac{36}{27}+\frac{15}{27}=\frac{51}{27}=\frac{17}{9}=1\frac{8}{9}$
10. $\frac{3}{12}+\frac{8}{12}=\frac{11}{12}$
    $\frac{11}{12}+\frac{7}{12}=\frac{18}{12}=\frac{3}{2}=1\frac{1}{2}$
11. $2\times 5{,}280=10{,}560$ ft
12. $7\times 5{,}280=36{,}960$ ft
13. $\frac{1}{2}\times 5{,}280=2{,}640$ ft
14. $\frac{7}{2}\times\frac{27}{4}=\frac{189}{8}=23\frac{5}{8}$ sq. ft
15. no
16. $8D=120$; $D=15$
    $\frac{1}{8}\cdot 8D=120\cdot\frac{1}{8}$
    $D=15$ years old

17. $12 \times 12 = 144$
18. $16 \div 2 = 8$
    $8 + 2 = 10$
    $10 \times 3 = 30$
    $30 \div 6 = 5$

Systematic Review 26E

1. $3Z - 13 = 20$
   $3Z = 33$
   $\frac{1}{3} \cdot 3Z = 33 \cdot \frac{1}{3}$
   $Z = 11$
2. $3(11) - 13 = 20$
   $33 - 13 = 20$
   $20 = 20$
3. $8L + 7 = 55$
   $8L = 48$
   $\frac{1}{8} \cdot 8L = 48 \cdot \frac{1}{8}$
   $L = 6$
4. $8(6) + 7 = 55$
   $48 + 7 = 55$
   $55 = 55$
5. $\frac{\cancel{22}^{11}}{7} \times \frac{1}{\cancel{2}} \times \frac{5}{12} = \frac{55}{84}$
6. $\frac{5}{8} \times \frac{\cancel{10}}{\cancel{3}} \times \frac{\cancel{21}^{7}}{\cancel{10}} = \frac{35}{8} = 4\frac{3}{8}$
7. $\frac{1}{3} \div \frac{5}{18} = \frac{1}{\cancel{3}} \times \frac{\cancel{18}^{6}}{5} = \frac{6}{5} = 1\frac{1}{5}$
8. $\frac{16}{3} \div \frac{3}{2} = \frac{16}{3} \times \frac{2}{3} = \frac{32}{9} = 3\frac{5}{9}$
9. $\frac{9}{15} > \frac{5}{15}$
10. $\frac{16}{32} = \frac{16}{32}$
11. $\frac{7}{21} < \frac{12}{21}$
12. $5 \times 5{,}280 = 26{,}400$ ft
13. $\frac{2}{3} \times \frac{5280}{1} = \frac{3520}{1} = 3{,}520$ ft
14. $\frac{7}{2} \times 16 = 56$ oz
15. $25\frac{4}{12} + 1\frac{9}{12} = 26\frac{13}{12} = 27\frac{1}{12}$ mi
16. 25: 1, <u>5</u>, 25
    45: 3, <u>5</u>, 9, 15, 45
    GCF=5
17. yes
18. $\frac{61}{10} \div \frac{3}{2} = \frac{61}{{}_{5}\cancel{10}} \times \frac{\cancel{2}}{3} = \frac{61}{15} = 4\frac{1}{15}$
    so 5 bags
19. $9 \times 9 = 81$
20. $35 \div 7 = 5$
    $5 + 4 = 9$
    $9 \times 6 = 54$
    $54 + 1 = 55$

Systematic Review 26F

1. $7A + 8 = 29$
   $7A = 21$
   $\frac{1}{7} \cdot 7A = 21 \cdot \frac{1}{7}$
   $A = 3$
2. $7(3) + 8 = 29$
   $21 + 8 = 29$
   $29 = 29$
3. $9M - 9 = 54$
   $9M = 63$
   $\frac{1}{9} \cdot 9M = 63 \cdot \frac{1}{9}$
   $M = 7$
4. $9(7) - 9 = 54$
   $63 - 9 = 54$
   $54 = 54$
5. $\frac{\cancel{6}}{\cancel{5}} \times \frac{3}{2} \times \frac{\cancel{5}}{\cancel{6}} = \frac{3}{2} = 1\frac{1}{2}$
6. $\frac{\cancel{3}}{\cancel{7}} \times \frac{5}{6} \times \frac{\cancel{7}}{\cancel{3}} = \frac{5}{6}$
7. $\frac{5}{8} \div \frac{1}{12} = \frac{5}{{}_{2}\cancel{8}} \times \frac{\cancel{12}^{3}}{1} = \frac{15}{2} = 7\frac{1}{2}$
8. $\frac{12}{5} \div \frac{36}{25} = \frac{\cancel{12}}{\cancel{5}} \times \frac{\cancel{25}^{5}}{{}_{3}\cancel{36}} = \frac{5}{3} = 1\frac{2}{3}$

9. $\frac{3}{4} - \frac{1}{4} = \frac{2}{4} = \frac{1}{2}$
10. $\frac{5}{10} - \frac{2}{10} = \frac{3}{10}$
11. $\frac{24}{30} - \frac{5}{30} = \frac{19}{30}$
12. $35 \div 3 = 11\frac{2}{3}$ yd
13. $8x2 = 16$ pt
14. $\frac{3}{2}x4 = 6$ qt
15. $10\frac{4}{8} - 6\frac{6}{8} = 9\frac{12}{8} - 6\frac{6}{8} = 3\frac{6}{8} = 3\frac{3}{4}$ tons

    $\frac{15}{\cancel{4}} x \frac{\overset{500}{\cancel{2000}}}{1} = 7{,}500$ lb
16. 3x3x7
17. $\frac{13}{\cancel{2}} x \overset{6}{\cancel{12}} = 78$ in
18. $\frac{7}{\cancel{8}} x \overset{660}{\cancel{5{,}280}} = 4{,}620$ ft

    $4{,}620 \div 3 = 1{,}540$ yd
19. $23x23 = 529$
20. $42 \div 6 = 7$

    $7 - 1 = 6$

    $6x6 = 36$

    $36 + 4 = 40$

Lesson Practice 27A

1. done
2. $\frac{22}{7}(3^2) = \frac{22}{7} x \frac{9}{1} = \frac{198}{7}$

   $= 28\frac{2}{7}$ sq. ft
3. $\frac{22}{7}(21^2) = \frac{22}{\cancel{7}} x \frac{\overset{63}{\cancel{441}}}{1} =$

   $\frac{1386}{1} = 1{,}386$ sq. in
4. $\frac{22}{7}(6^2) = \frac{22}{7} x \frac{36}{1} = \frac{792}{7}$

   $= 113\frac{1}{7}$ sq. ft
5. $\frac{22}{7}(4^2) = \frac{22}{7} x \frac{16}{1} = \frac{352}{7}$

   $= 50\frac{2}{7}$ sq. yd
6. done
7. $\frac{2}{1} x \frac{22}{7} x \frac{3}{1} = \frac{132}{7} = 18\frac{6}{7}$ ft
8. $\frac{2}{1} x \frac{22}{\cancel{7}} x \frac{\overset{3}{\cancel{21}}}{1} = \frac{132}{1} = 132$ in
9. $\frac{2}{1} x \frac{22}{7} x \frac{6}{1} = \frac{264}{7} = 37\frac{5}{7}$ ft
10. $\frac{2}{1} x \frac{22}{\cancel{7}} x \frac{\cancel{7}}{1} = \frac{44}{1} = 44$ in

Lesson Practice 27B

1. $\frac{22}{7}(4^2) = \frac{22}{7} x \frac{16}{1} = \frac{352}{7}$

   $= 50\frac{2}{7}$ sq. in
2. $\frac{22}{7}(28^2) = \frac{22}{\cancel{7}} x \frac{\overset{112}{\cancel{784}}}{1} = \frac{2464}{1}$

   $= 2{,}464$ sq. ft
3. $\frac{22}{7}(11^2) = \frac{22}{7} x \frac{121}{1} = \frac{2662}{7}$

   $= 380\frac{2}{7}$ sq. in
4. $\frac{22}{7}(70^2) = \frac{22}{\cancel{7}} x \frac{\overset{700}{\cancel{4900}}}{1} = \frac{15400}{1}$

   $= 15{,}400$ sq. ft
5. $\frac{22}{7}(10^2) = \frac{22}{7} x \frac{100}{1} = \frac{2200}{7}$

   $= 314\frac{2}{7}$ sq. ft
6. $\frac{2}{1} x \frac{22}{7} x \frac{4}{1} = \frac{176}{7} = 25\frac{1}{7}$ in
7. $\frac{2}{1} x \frac{22}{\cancel{7}} x \frac{\overset{4}{\cancel{28}}}{1} = \frac{176}{1} = 176$ ft
8. $\frac{2}{1} x \frac{22}{7} x \frac{11}{1} = \frac{484}{7} = 69\frac{1}{7}$ in
9. $\frac{2}{1} x \frac{22}{\cancel{7}} x \frac{\overset{10}{\cancel{70}}}{1} = \frac{440}{1} = 440$ ft

10. $\frac{2}{1} \times \frac{22}{7} \times \frac{9}{1} = \frac{396}{7} = 56\frac{4}{7}$ in

Lesson Practice 27C

1. $\frac{22}{7}\left(49^2\right) = \frac{22}{\cancel{7}} \times \frac{\overset{343}{\cancel{2401}}}{1} =$

   $\frac{7546}{1} = 7{,}546$ sq. in

2. $\frac{22}{7}\left(5^2\right) = \frac{22}{7} \times \frac{25}{1} = \frac{550}{7}$

   $= 78\frac{4}{7}$ sq. ft

3. $\frac{22}{7}\left(8^2\right) = \frac{22}{7} \times \frac{64}{1} = \frac{1408}{7}$

   $= 201\frac{1}{7}$ sq. in

4. $\frac{22}{7}\left(42^2\right) = \frac{22}{\cancel{7}} \times \frac{\overset{252}{\cancel{1764}}}{1} =$

   $\frac{5544}{1} = 5{,}544$ sq. ft

5. $\frac{22}{7}\left(12^2\right) = \frac{22}{7} \times \frac{144}{1} = \frac{3168}{7}$

   $= 452\frac{4}{7}$ sq. ft

6. $\frac{2}{1} \times \frac{22}{\cancel{7}} \times \frac{\overset{7}{\cancel{49}}}{1} = \frac{308}{1} = 308$ in

7. $\frac{2}{1} \times \frac{22}{7} \times \frac{5}{1} = \frac{220}{7} = 31\frac{3}{7}$ ft

8. $\frac{2}{1} \times \frac{22}{7} \times \frac{8}{1} = \frac{352}{7} = 50\frac{2}{7}$ in

9. $\frac{2}{1} \times \frac{22}{\cancel{7}} \times \frac{\overset{6}{\cancel{42}}}{1} = \frac{264}{1} = 264$ ft

10. $\frac{2}{1} \times \frac{22}{\cancel{7}} \times \frac{\overset{5}{\cancel{35}}}{1} = \frac{220}{1} = 220$ mi

Systematic Review 27D

1. $\frac{22}{7}\left(7^2\right) = \frac{22}{\cancel{7}} \times \frac{\overset{7}{\cancel{49}}}{1} = \frac{154}{1} =$

   154 sq. in

2. $\frac{2}{1} \times \frac{22}{\cancel{7}} \times \frac{\cancel{7}}{1} = \frac{44}{1} = 44$ in

3. $\frac{22}{7}\left(9^2\right) = \frac{22}{7} \times \frac{81}{1} = \frac{1782}{7} =$

   $254\frac{4}{7}$ sq. ft

4. $\frac{2}{1} \times \frac{22}{7} \times \frac{9}{1} = \frac{396}{7} = 56\frac{4}{7}$ ft

5. $3P - 15 = 12$

   $3P = 27$

   $\frac{1}{3} \cdot 3P = 27 \cdot \frac{1}{3}$

   $P = 9$

6. $3(9) - 15 = 12$

   $27 - 15 = 12$

   $12 = 12$

7. $\frac{\cancel{6}}{\cancel{5}} \times \frac{\cancel{5}}{\underset{2}{\cancel{12}}} \times \frac{3}{7} = \frac{3}{14}$

8. $\frac{3}{\cancel{5}} \times \frac{\overset{4}{\cancel{20}}}{\underset{2}{\cancel{14}}} \times \frac{\cancel{7}}{2} = \frac{12}{4} = 3$

9. $\frac{9}{11} \div \frac{1}{11} = \frac{9 \div 1}{1} = 9$

10. $\frac{9}{4} \div \frac{27}{22} = \frac{\cancel{9}}{\underset{2}{\cancel{4}}} \times \frac{\overset{11}{\cancel{22}}}{\underset{3}{\cancel{27}}} = \frac{11}{6} = 1\frac{5}{6}$

11. done

12. 8 pt = 8 pt

13. 8 oz > 7 oz

14. $\frac{22}{7}\left(16^2\right) = \frac{22}{7} \times \frac{256}{1} = \frac{5632}{7} =$

    $804\frac{4}{7}$ sq. in

15. $\frac{1}{2} \times \frac{1}{2} \times \frac{1}{2} = \frac{1}{8}$ cu. ft

16. $\frac{1}{2} \times 12 = 6$ in; $6 \times 6 \times 6 = 216$ cu. in

17. $2\frac{1}{4} + 5\frac{1}{4} = 7\frac{2}{4} = 7\frac{1}{2}$

    $7\frac{1}{2} + 1\frac{1}{2} = 8\frac{2}{2} = 9$ in

18. $10 \times 10 = 100$
    $100 \div 2 = 50$
    $50 + 3 = 53$
    $53 - 4 = 49$

Systematic Review 27E

1. $\frac{22}{7}\left(56^2\right) = \frac{22}{\cancel{7}} \times \frac{\overset{448}{\cancel{3136}}}{1}$
   $= 9856$ sq. in
2. $\frac{2}{1} \times \frac{22}{\cancel{7}} \times \frac{\overset{8}{\cancel{56}}}{1} = \frac{352}{1} = 352$ in
3. $\frac{22}{7}\left(6^2\right) = \frac{22}{7} \times \frac{36}{1} = \frac{792}{7}$
   $= 113\frac{1}{7}$ sq. ft
4. $\frac{2}{1} \times \frac{22}{7} \times \frac{6}{1} = \frac{264}{7} = 37\frac{5}{7}$ ft
5. $9D + 14 = 23$
   $9D = 9$
   $\frac{1}{9} \cdot 9D = 9 \cdot \frac{1}{9}$
   $D = 1$
6. $9(1) + 14 = 23$
   $9 + 14 = 23$
   $23 = 23$
7. $\frac{\cancel{9}}{\cancel{4}} \times \frac{\overset{2}{\cancel{8}}}{3} \times \frac{2}{\cancel{9}} = \frac{4}{3} = 1\frac{1}{3}$
8. $\frac{5}{\underset{\cancel{2}}{\cancel{6}}} \times \frac{\overset{\cancel{3}}{\cancel{9}}}{\cancel{2}} \times \frac{\overset{\cancel{2}}{\cancel{4}}}{\cancel{3}} = \frac{5}{1} = 5$
9. $\frac{4}{7} \div \frac{7}{3} = \frac{4}{7} \times \frac{3}{7} = \frac{12}{49}$
10. $\frac{32}{3} \div \frac{17}{4} = \frac{32}{3} \times \frac{4}{17}$
    $= \frac{128}{51} = 2\frac{26}{51}$
11. 15 qt < 32 qt
12. 4,000 lb > 1,500 lb
13. 1,760 ft = 1,760 ft
14. $\frac{3}{4} \times 2{,}000 = 1{,}500$ lb
15. $\frac{22}{7}\left(14^2\right) = \frac{22}{\cancel{7}} \times \frac{\overset{28}{\cancel{196}}}{1} = \frac{616}{1}$
    $= 616$ sq ft
16. $\frac{2}{1} \times \frac{22}{\cancel{7}} \times \frac{\overset{2}{\cancel{14}}}{1} = \frac{88}{1} = 88$ ft
17. $\frac{1}{2} \times \frac{9}{\cancel{2}} \times \frac{\overset{4}{\cancel{8}}}{1} =$
    $\frac{1}{\cancel{2}} \times \frac{9}{1} \times \frac{\overset{2}{\cancel{4}}}{1} = \frac{18}{1} = 18$ sq. in
18. $\frac{21}{2} \div \frac{7}{6} = \frac{\overset{3}{\cancel{21}}}{\cancel{2}} \times \frac{\overset{3}{\cancel{6}}}{\cancel{7}}$
    $= \frac{9}{1} = 9$ cakes
19. $3A = 90$
    $\frac{1}{3} \cdot 3A = 90 \cdot \frac{1}{3}$
    $A = 30$ years old
20. $11 \times 3 = 33$
    $33 - 1 = 32$
    $32 \div 8 = 4$
    $4 \times 7 = 28$

Systematic Review 27F

1. $\frac{22}{7}\left(21^2\right) = \frac{22}{\cancel{7}} \times \frac{\overset{63}{\cancel{441}}}{1} = \frac{1386}{1}$
   $= 1{,}386$ sq. in
2. $\frac{2}{1} \times \frac{22}{\cancel{7}} \times \frac{\overset{3}{\cancel{21}}}{1} = \frac{132}{1} = 132$ in
3. $\frac{22}{7}\left(10^2\right) = \frac{22}{7} \times \frac{100}{1} = \frac{2200}{7}$
   $= 314\frac{2}{7}$ sq. ft
4. $\frac{2}{1} \times \frac{22}{7} \times \frac{10}{1} = \frac{440}{7} = 62\frac{6}{7}$ ft

5. $4Q - 5 = 3$
$\quad +5 \quad +5$
$4Q = 8$
$\frac{1}{4} \cdot 4Q = \frac{1}{4} \cdot 8$
$Q = 2$

6. $4(2) - 5 = 3$
$8 - 5 = 3$
$3 = 3$

7. $\frac{\overset{3}{\cancel{12}}}{5} \times \frac{11}{\cancel{4}} = \frac{33}{5} = 6\frac{3}{5}$

8. $\frac{\overset{9}{\cancel{45}}}{\cancel{8}} \times \frac{\cancel{8}}{\cancel{5}} \times \frac{15}{7} = \frac{135}{7} = 19\frac{2}{7}$

9. $\frac{4}{10} \div \frac{2}{5} = \frac{\overset{2}{\cancel{4}}}{\underset{2}{\cancel{10}}} \times \frac{\cancel{5}}{\cancel{2}} = \frac{1}{1} = 1$

10. $\frac{66}{7} \div \frac{21}{10} = \frac{\overset{22}{\cancel{66}}}{7} \times \frac{10}{\underset{7}{\cancel{21}}}$
$= \frac{220}{49} = 4\frac{24}{49}$

11. 204 in > 17 in

12. 10,000 lb = 10,000 lb

13. 3,960 ft < 4,000 ft

14. $\frac{22}{7}(1^2) = \frac{22}{7} \times \frac{1}{1} = \frac{22}{7}$
$= 3\frac{1}{7}$ sq. mi

15. $\frac{22}{7}(2^2) = \frac{22}{7} \times \frac{4}{1} = \frac{88}{7}$
$= 12\frac{4}{7}$ sq. mi; no

16. $10 - 6\frac{3}{4} = 3\frac{1}{4}$ lb

17. $\frac{1}{2} \times \frac{9}{1} \times \frac{10}{3} = \frac{15}{1} = 15$ sq. in

18. $20\frac{1}{3} + 20\frac{1}{3} = 40\frac{2}{3}$
$30\frac{1}{2} + 30\frac{1}{2} = 60\frac{2}{2} = 61$
$40\frac{2}{3} + 61 = 101\frac{2}{3}$ ft

19. $101\frac{2}{3} \times \frac{3}{5} = \frac{\overset{61}{\cancel{305}}}{\cancel{3}} \times \frac{\cancel{3}}{\cancel{5}} = 61$ ft

20. $12 + 8 = 20$
$20 - 10 = 10$
$10 \times 4 = 40$
$40 + 7 = 47$

Lesson Practice 28A

1. done
2. done
Feel free to use canceling to simplify your work.

3. $\frac{5}{1} \cdot \frac{1}{5}A = \frac{3}{1} \cdot \frac{5}{1}$
$A = \frac{15}{1} = 15$

4. $\frac{1}{5}(15) = 3$
$3 = 3$

5. $\frac{1}{7}Y - 4 = 2$
$\frac{1}{7}Y = 6$
$\frac{7}{1} \cdot \frac{1}{7}Y = \frac{6}{1} \cdot \frac{7}{1}$
$Y = \frac{42}{1} = 42$

6. $\frac{1}{7}(42) - 4 = 2$
$6 - 4 = 2$
$2 = 2$

7. $\frac{3}{8}B + 8 = 14$
$\frac{3}{8}B = 6$
$\frac{8}{3} \cdot \frac{3}{8}B = \frac{6}{1} \cdot \frac{8}{3}$
$B = \frac{48}{3} = 16$

8. $\frac{3}{8} \cdot \frac{16}{1} + 8 = 14$
$\frac{48}{8} + 8 = 14$
$6 + 8 = 14$
$14 = 14$

9. $\frac{3}{2} \cdot \frac{2}{3}Z = \frac{12}{1} \cdot \frac{3}{2}$

$Z = \frac{36}{2} = 18$

10. $\frac{2}{3} \cdot \frac{18}{1} = 12$

$\frac{36}{3} = 12$

$12 = 12$

11. $\frac{2}{5}C - 2 = 2$

$\frac{2}{5}C = 4$

$\frac{5}{2} \cdot \frac{2}{5}C = \frac{4}{1} \cdot \frac{5}{2}$

$C = \frac{20}{2} = 10$

12. $\frac{2}{5} \cdot \frac{10}{1} - 2 = 2$

$\frac{20}{5} - 2 = 2$

$4 - 2 = 2$

$2 = 2$

Lesson Practice 28B

1. $\frac{1}{4}G - 4 = 2$

$\frac{1}{4}G = 6$

$\frac{4}{1} \cdot \frac{1}{4}G = \frac{6}{1} \cdot \frac{4}{1}$

$G = \frac{24}{1} = 24$

2. $\frac{1}{4}(24) - 4 = 2$

$\frac{1}{4} \cdot \frac{24}{1} - 4 = 2$

$\frac{24}{4} - 4 = 2$

$6 - 4 = 2$

$2 = 2$

3. $\frac{5}{4} \cdot \frac{4}{5}D = \frac{40}{1} \cdot \frac{5}{4}$

$D = \frac{200}{4} = 50$

4. $\frac{4}{5} \cdot \frac{50}{1} = 40$

$\frac{200}{5} = 40$

$40 = 40$

5. $\frac{1}{10}H - 4 = 0$

$\frac{1}{10}H = 4$

$\frac{10}{1} \cdot \frac{1}{10}H = \frac{4}{1} \cdot \frac{10}{1}$

$H = \frac{40}{1} = 40$

6. $\frac{1}{10}(40) - 4 = 0$

$4 - 4 = 0$

$0 = 0$

7. $\frac{5}{8}E - 5 = 20$

$\frac{5}{8}E = 25$

$\frac{8}{5} \cdot \frac{5}{8}E = \frac{25}{1} \cdot \frac{8}{5}$

$E = \frac{200}{5} = 40$

8. $\frac{5}{8}(40) - 5 = 20$

$\frac{5}{8} \cdot \frac{40}{1} - 5 = 20$

$\frac{200}{8} - 5 = 20$

$25 - 5 = 20$

9. $\frac{4}{1} \cdot \frac{1}{4}J = \frac{7}{1} \cdot \frac{4}{1}$

$J = \frac{28}{1} = 28$

10. $\frac{1}{4}(28) = 7$

$7 = 7$

11. $\frac{5}{9}F + 9 = 14$

$\frac{5}{9}F = 5$

$\frac{9}{5} \cdot \frac{5}{9}F = \frac{5}{1} \cdot \frac{9}{5}$

$F = \frac{45}{5} = 9$

12. $\frac{\cancel{5}}{9} \cdot \frac{45}{\cancel{5}} + 9 = 14$

$\frac{45}{9} + 9 = 14$

$5 + 9 = 14$

$14 = 14$

Lesson Practice 28C

1. $\frac{1}{3}K + 3 = 11$

$\frac{1}{3}K = 8$

$\frac{3}{1} \cdot \frac{1}{3}K = \frac{8}{1} \cdot \frac{3}{1}$

$K = \frac{24}{1} = 24$

2. $\frac{1}{3} \cdot \frac{24}{1} + 3 = 11$

$\frac{24}{3} + 3 = 11$

$8 + 3 = 11$

$11 = 11$

3. $\frac{5}{1} \cdot \frac{1}{5}G = \frac{8}{1} \cdot \frac{5}{1}$

$G = \frac{40}{1} = 40$

4. $\frac{1}{5} \cdot \frac{40}{1} = 8$

$\frac{40}{5} = 8$

$8 = 8$

5. $\frac{3}{5}L + 3 = 18$

$\frac{3}{5}L = 15$

$\frac{5}{3} \cdot \frac{3}{5}L = \frac{15}{1} \cdot \frac{5}{3}$

$L = \frac{75}{3} = 25$

6. $\frac{3}{5}(25) + 3 = 18$

$\frac{3}{5} \cdot \frac{25}{1} + 3 = 18$

$\frac{75}{5} + 3 = 18$

$15 + 3 = 18$

7. $\frac{5}{12}R + 5 = 25$

$\frac{5}{12}R = 20$

$\frac{12}{5} \cdot \frac{5}{12}R = \frac{20}{1} \cdot \frac{12}{5}$

$R = \frac{240}{5} = 48$

8. $\frac{5}{12} \cdot \frac{48}{1} + 5 = 25$

$\frac{240}{12} + 5 = 25$

$20 + 5 = 25$

$25 = 25$

9. $\frac{3}{2} \cdot \frac{2}{3}M = \frac{10}{1} \cdot \frac{3}{2}$

$M = \frac{30}{2} = 15$

10. $\frac{2}{3} \cdot \frac{15}{1} = 10$

$\frac{30}{3} = 10$

$10 = 10$

11. $\frac{2}{9}S + 6 = 14$

$\frac{2}{9}S = 8$

$\frac{9}{2} \cdot \frac{2}{9}S = \frac{8}{1} \cdot \frac{9}{2}$

$S = \frac{72}{2} = 36$

12. $\frac{2}{9}(36) + 6 = 14$

$\frac{2}{9} \cdot \frac{36}{1} + 6 = 14$

$\frac{72}{9} + 6 = 14$

$8 + 6 = 14$

$14 = 14$

Systematic Review 28D

1. $\frac{1}{5}T+4=8$

   $\frac{1}{5}T=4$

   $\frac{5}{1}\cdot\frac{1}{5}T=\frac{4}{1}\cdot\frac{5}{1}$

   $T=\frac{20}{1}=20$

2. $\frac{1}{5}(20)+4=8$

   $\frac{1}{5}\cdot\frac{20}{1}+4=8$

   $\frac{20}{5}+4=8$

   $4+4=8$

   $8=8$

3. $\frac{3}{4}N-6=21$

   $\frac{3}{4}N=27$

   $\frac{4}{3}\cdot\frac{3}{4}N=\frac{27}{1}\cdot\frac{4}{3}$

   $N=\frac{108}{3}$

   $N=36$

4. $\frac{3}{4}(36)-6=21$

   $\frac{3}{4}\cdot\frac{36}{1}-6=21$

   $\frac{108}{4}-6=21$

   $27-6=21$

   $21=21$

5. $\frac{22}{7}\left(14^2\right)=\frac{22}{\cancel{7}}\cdot\frac{\overset{28}{\cancel{196}}}{1}=$

   $\frac{616}{1}=616$ sq. ft

6. $\frac{2}{1}\cdot\frac{22}{\cancel{7}}\cdot\frac{\overset{2}{\cancel{14}}}{1}=\frac{88}{1}=88$ ft

7. $\frac{1}{\cancel{2}}\times\frac{\cancel{2}}{\cancel{3}}\times\frac{\overset{3}{\cancel{9}}}{11}=\frac{3}{11}$

8. $\frac{\cancel{7}}{\cancel{8}}\times\frac{\cancel{8}}{\cancel{3}}\times\frac{\overset{2}{\cancel{6}}}{\cancel{7}}=\frac{2}{1}=2$

9. $\frac{3}{1}\div\frac{1}{2}=\frac{3}{1}\times\frac{2}{1}=\frac{6}{1}=6$

10. $\frac{31}{5}\div\frac{31}{20}=\frac{\cancel{31}}{\cancel{5}}\times\frac{\overset{4}{\cancel{20}}}{\cancel{31}}=\frac{4}{1}=4$

11. $21\text{ ft}<36\text{ ft}$

12. $10\text{ pt}<11\text{ pt}$

13. $5\text{ qt}>4\text{ qt}$

14. $2+7+9=18$

    $18\div3=6$

15. $5+5+9+13=32$

    $32\div4=8$

16. $2+5+8+9=24$

    $24\div4=6$

17. $77+80+95+100=352$

    $352\div4=88$ average score

18. $2+7=9$

    $9\times7=63$

    $63+2=65$

    $65+5=70$

Systematic Review 28E

1. $\frac{1}{6}U+2=3$

   $\frac{1}{6}U=1$

   $\frac{6}{1}\cdot\frac{1}{6}U=\frac{1}{1}\cdot\frac{6}{1}$

   $U=\frac{6}{1}=6$

2. $\frac{1}{6}(6)+2=3$

   $\frac{1}{6}\cdot\frac{6}{1}+2=3$

   $1+2=3$

   $3=3$

3. $\frac{5}{2}\cdot\frac{2}{5}P=\frac{2}{1}\cdot\frac{5}{2}$

   $P=\frac{10}{2}=5$

4. $\frac{2}{5}\cdot\frac{5}{1}=2$

   $\frac{10}{5}=2$

   $2=2$

5. $\frac{22}{7}\left(42^2\right) = \frac{22}{\cancel{7}} \cdot \frac{\overset{252}{\cancel{1764}}}{1} =$

$\frac{5544}{1} = 5{,}544$ sq. in

6. $\frac{2}{1} \times \frac{22}{\cancel{7}} \times \frac{\overset{6}{\cancel{42}}}{1} = 264$ in

7. $\frac{\overset{3}{\cancel{15}}}{\underset{2}{\cancel{14}}} \times \frac{\cancel{7}}{4} \times \frac{1}{\underset{5}{\cancel{25}}} = \frac{3}{40}$

8. $\frac{\overset{2}{\cancel{4}}}{\underset{3}{\cancel{9}}} \times \frac{\cancel{5}}{\cancel{2}} \times \frac{\cancel{3}}{\cancel{5}} = \frac{2}{3}$

9. $\frac{3}{6} \div \frac{2}{3} = \frac{3}{6} \times \frac{3}{2} = \frac{9}{12} = \frac{3}{4}$

10. $\frac{24}{7} \div \frac{18}{7} = \frac{24 \div 18}{1} =$

$\frac{24}{18} = \frac{4}{3} = 1\frac{1}{3}$

11. 32 oz = 32 oz

12. 14,000 lb > 8,000 lb

13. 120 in < 125 in

14. $4+5+6=15$

$15 \div 3 = 5$

15. $6+7+9+10=32$

$32 \div 4 = 8$

16. $4+7+10=21$

$21 \div 3 = 7$

17. $4\frac{1}{2} + 4\frac{1}{2} = 8\frac{2}{2} = 9$

$6\frac{1}{3} + 6\frac{1}{3} = 12\frac{2}{3}$

$12\frac{2}{3} + 9 = 21\frac{2}{3}$ in

18. $\frac{\overset{3}{\cancel{9}}}{2} \times \frac{19}{\cancel{3}} = \frac{57}{2} =$

$28\frac{1}{2}$ sq. in

19. 2x2x2x2x2x2

20. $3 \times 2 = 6$

$6 \times 3 = 18$

$18 + 2 = 20$

$20 - 4 = 16$

Systematic Review 28F

1. $\frac{3}{16}V - 3 = 12$

$\frac{3}{16}V = 15$

$\frac{16}{3} \cdot \frac{3}{16}V = \frac{15}{1} \cdot \frac{16}{3}$

$V = \frac{240}{3} = 80$

2. $\frac{3}{16}(80) - 3 = 12$

$\frac{3}{16} \cdot \frac{80}{1} - 3 = 12$

$\frac{240}{16} - 3 = 12$

$15 - 3 = 12$

$12 = 12$

3. $5Q + 4 = 19$

$5Q = 15$

$\frac{1}{5} \cdot 5Q = \frac{15}{1} \cdot \frac{1}{5}$

$Q = \frac{15}{5} = 3$

4. $5(3) + 4 = 19$

$15 + 4 = 19$

$19 = 19$

5. $\frac{22}{7}\left(2^2\right) = \frac{22}{7} \cdot \frac{4}{1} =$

$\frac{88}{7} = 12\frac{4}{7}$ sq. in

6. $\frac{2}{1} \cdot \frac{22}{7} \cdot \frac{2}{1} =$

$\frac{88}{7} = 12\frac{4}{7}$ in

7. $\frac{\cancel{2}}{1} \times \frac{\cancel{3}}{\underset{\cancel{5}}{\cancel{10}}} \times \frac{\cancel{5}}{\underset{2}{\cancel{6}}} = \frac{1}{2}$

8. $\frac{\overset{2}{\cancel{10}}}{\underset{\cancel{3}}{\cancel{21}}} \times \frac{\overset{2}{\cancel{14}}}{\cancel{5}} \times \frac{\overset{2}{\cancel{6}}}{11} = \frac{8}{11}$

9. $\frac{1}{2} \div \frac{1}{10} = \frac{1}{2} \times \frac{10}{1} = \frac{10}{2} = 5$

10. $\frac{37}{8} \div \frac{41}{12} = \frac{37}{\underset{2}{\cancel{8}}} \times \frac{\overset{3}{\cancel{12}}}{41} =$

$\frac{111}{82} = 1\frac{29}{82}$

11. 256 oz > 1 oz

12. 2,640 ft = 2,640 ft

13. 9 in < 10 in
14. $4+8+10+14=36$
    $36 \div 4 = 9$
15. $6+7+9+14=36$
    $36 \div 4 = 9$
16. $6+6+12=24$
    $24 \div 3 = 8$
17. $2\frac{1}{4}+1\frac{1}{4}=3\frac{2}{4}=3\frac{1}{2}$
    $3\frac{1}{2}+1\frac{1}{2}=4\frac{2}{2}=5$
    $5 \div 3 = \frac{5}{3} = 1\frac{2}{3}$ mi
18. $\frac{4}{3} \times \frac{4}{3} \times \frac{4}{3} = \frac{64}{27} = 2\frac{10}{27}$ cu. in
19. 30 : 2, 3, 5, 6, 10, 15, 30
    45 : 3, 5, 9, 15, 45
    GCF=15
20. $20 \div 5 = 4$
    $4 \times 3 = 12$
    $12+7=19$
    $19-4=15$

Lesson Practice 29A

1. done
2. $\frac{1}{2} = \frac{5}{10} = .5 = \frac{50}{100} = .50$
3. $\frac{4}{5} = \frac{8}{10} = .8 = \frac{80}{100} = .80$
4. done
5. $\frac{5}{5} = \frac{100}{100} = 100\%$
6. done
7. $\frac{1}{2} = \frac{50}{100} = .50 = 50\%$

Lesson Practice 29B

1. done
2. $\frac{3}{4} = \frac{75}{100} = .75$
3. $\frac{2}{5} = \frac{4}{10} = .4 = \frac{40}{100} = .40$
4. $\frac{3}{5} = \frac{60}{100} = 60\%$
5. $\frac{4}{5} = \frac{80}{100} = 80\%$
6. $\frac{1}{4} = \frac{25}{100} = .25 = 25\%$
7. $\frac{3}{4} = \frac{75}{100} = .75 = 75\%$

Lesson Practice 29C

1. $\frac{1}{5} = \frac{2}{10} = .2 = \frac{20}{100} = .20$
2. $\frac{3}{5} = \frac{6}{10} = .6 = \frac{60}{100} = .60$
3. $\frac{1}{4} = \frac{25}{100} = .25$
4. $\frac{5}{5} = \frac{100}{100} = 100\%$
5. $\frac{2}{5} = \frac{40}{100} = 40\%$
6. $\frac{3}{4} = \frac{75}{100} = .75 = 75\%$
7. $\frac{1}{2} = \frac{50}{100} = .50 = 50\%$

Systematic Review 29D

1. $\frac{1}{5} = \frac{20}{100} = 20\%$
2. $\frac{3}{5} = \frac{60}{100} = 60\%$
3. $\frac{1}{4} = \frac{25}{100} = .25 = 25\%$
4. $\frac{4}{5} = \frac{80}{100} = .80 = 80\%$
5. $\frac{1}{2}T+5=8$
   $\frac{1}{2}T=3$
   $\frac{2}{1} \cdot \frac{1}{2}T = \frac{3}{1} \cdot \frac{2}{1}$
   $T = \frac{6}{1} = 6$

6. $\frac{1}{2}(6)+5=8$

   $3+5=8$

   $8=8$

7. $\frac{\cancel{3}}{\cancel{4}_{2}}\times\frac{\cancel{2}}{7}\times\frac{1}{\cancel{3}}=\frac{1}{14}$

8. $\frac{\overset{3}{\cancel{\overset{21}{\cancel{42}}}}}{\underset{2}{\cancel{\underset{4}{\cancel{8}}}}}\times\frac{47}{\cancel{6}}\times\frac{\overset{2}{\cancel{12}}}{\cancel{7}}=\frac{141}{2}=70\frac{1}{2}$

9. 6
10. 13
11. 9
12. 31
13. XX
14. VIII
15. XVI
16. XXXIV
17. $16+18+22=56$

    $56\div3=18\frac{2}{3}$ is average age

18. $A=\frac{22}{7}\left(70^2\right)=\frac{22}{\cancel{7}}\cdot\frac{\overset{700}{\cancel{4900}}}{1}=$

    $\frac{15400}{1}=15{,}400$ sq. ft

    $C=\frac{2}{1}\cdot\frac{22}{\cancel{7}}\cdot\frac{\overset{10}{\cancel{70}}}{1}=\frac{440}{1}=440$ ft

Systematic Review 29E

1. $\frac{2}{5}=\frac{40}{100}=40\%$
2. $\frac{5}{5}=\frac{100}{100}=100\%$
3. $\frac{3}{4}=\frac{75}{100}=.75=75\%$
4. $\frac{1}{2}=\frac{50}{100}=.50=50\%$
5. $\frac{4}{5}U-11=33$

   $\frac{4}{5}U=44$

   $\frac{5}{4}\cdot\frac{4}{5}U=\frac{44}{1}\cdot\frac{5}{4}$

   $U=\frac{220}{4}=55$

6. $\frac{4}{5}(55)-11=33$

   $\frac{4}{5}\cdot\frac{55}{1}-11=33$

   $\frac{220}{5}-11=33$

   $44-11=33;\ 33=33$

7. $\frac{9}{12}+\frac{8}{12}+\frac{1}{12}=\frac{18}{12}=\frac{3}{2}=1\frac{1}{2}$
8. $\frac{5}{15}+\frac{6}{15}+\frac{1}{15}=\frac{12}{15}=\frac{4}{5}$
9. 2
10. 17
11. 27
12. 19
13. VII
14. XVIII
15. XXI
16. XXXV
17. $(\frac{1}{2})(24)(\frac{13}{2})=78$ sq. in
18. $\frac{3}{8}\div\frac{5}{8}=\frac{3}{5}$ mi
19. $3\frac{2}{6}+6\frac{3}{6}+5\frac{1}{6}=14\frac{6}{6}=15$

    $15\div3=5$ in

20. $\frac{2}{1}\cdot\frac{22}{\cancel{7}}\cdot\frac{\cancel{7}}{1}=\frac{44}{1}=44$ ft

Systematic Review 29F

1. $\frac{3}{5}=\frac{60}{100}=60\%$
2. $\frac{1}{5}=\frac{20}{100}=20\%$
3. $\frac{4}{5}=\frac{80}{100}=.80=80\%$
4. $\frac{1}{4}=\frac{25}{100}=.25=25\%$

5. $7X - 9 = 19$
$7X = 28$
$\frac{1}{7} \cdot 7X = 28 \cdot \frac{1}{7}$
$X = 4$

6. $7(4) - 9 = 19$
$28 - 9 = 19$
$19 = 19$

7. $16\frac{5}{9} - 8\frac{1}{9} = 8\frac{4}{9}$

8. $6\frac{5}{5} - 3\frac{4}{5} = 3\frac{1}{5}$

9. $4\frac{15}{40} + 2\frac{24}{40} = 6\frac{39}{40}$

10. 3

11. 11

12. 25

13. 39

14. V

15. XII

16. XXXVIII

17. XIV

18. $45 + 34 + 40 = 119$
$119 \div 3 = 39\frac{2}{3}$ average miles

19. $\frac{1}{2} \times \frac{1}{\cancel{4}} \times \frac{\cancel{4}}{5} = \frac{1}{10}$ of the job

20. $\frac{22}{7}\left(28^2\right) = \frac{22}{\cancel{7}} \cdot \frac{\overset{112}{\cancel{784}}}{1} =$
$\frac{2464}{1} = 2{,}464$ sq ft
$2{,}464 > 200$; no

Lesson Practice 30A

1. done

2. done

3. $\frac{3}{5}X - \frac{1}{2} = 19\frac{1}{2}$
$\frac{3}{5}X = 19\frac{1}{2} + \frac{1}{2}$
$\frac{3}{5}X = 20$
$\frac{5}{3} \cdot \frac{3}{5}X = \frac{20}{1} \cdot \frac{5}{3}$
$X = \frac{100}{3} = 33\frac{1}{3}$

4. $\frac{3}{5} \cdot \frac{100}{3} - \frac{1}{2} = 19\frac{1}{2}$
$\frac{300}{15} - \frac{1}{2} = 19\frac{1}{2}$
$20 - \frac{1}{2} = 19\frac{1}{2}$
$19\frac{1}{2} = 19\frac{1}{2}$

5. $\frac{1}{6}L + \frac{1}{3} = \frac{2}{3}$
$\frac{1}{6}L = \frac{1}{3}$
$\frac{6}{1} \cdot \frac{1}{6}L = \frac{1}{3} \cdot \frac{6}{1}$
$L = \frac{6}{3} = 2$

6. $\frac{1}{6}(2) + \frac{1}{3} = \frac{2}{3}$
$\frac{1}{6} \cdot \frac{2}{1} + \frac{1}{3} = \frac{2}{3}$
$\frac{2}{6} + \frac{1}{3} = \frac{2}{3}$
$\frac{1}{3} + \frac{1}{3} = \frac{2}{3}$
$\frac{2}{3} = \frac{2}{3}$

7. $\frac{1}{2}Y + \frac{3}{4} = 7\frac{3}{4}$
$\frac{1}{2}Y = 7$
$\frac{2}{1} \cdot \frac{1}{2}Y = \frac{7}{1} \cdot \frac{2}{1}$
$Y = \frac{14}{1} = 14$

8. $\frac{1}{2}(14)+\frac{3}{4}=7\frac{3}{4}$

$7+\frac{3}{4}=7\frac{3}{4}$

$7\frac{3}{4}=7\frac{3}{4}$

9. $\frac{4}{5}M-\frac{1}{10}=\frac{6}{10}$

$\frac{4}{5}M=\frac{7}{10}$

$\frac{5}{4}\cdot\frac{4}{5}M=\frac{7}{\cancel{10}_2}\cdot\frac{\cancel{5}}{4}$

$M=\frac{7}{8}$

10. $\frac{4}{5}\cdot\frac{7}{8}-\frac{1}{10}=\frac{3}{5}$

$\frac{28}{40}-\frac{1}{10}=\frac{3}{5}$

$\frac{7}{10}-\frac{1}{10}=\frac{3}{5}$

$\frac{6}{10}=\frac{3}{5}$

11. $\frac{1}{4}Z-\frac{2}{3}=3\frac{1}{3}$

$\frac{1}{4}Z=3\frac{3}{3}=4$

$\frac{4}{1}\cdot\frac{1}{4}Z=4\cdot\frac{4}{1}$

$Z=16$

12. $\frac{1}{4}(16)-\frac{2}{3}=3\frac{1}{3}$

$4-\frac{2}{3}=3\frac{1}{3}$

$3\frac{1}{3}=3\frac{1}{3}$

Lesson Practice 30B

1. $\frac{1}{2}P-\frac{1}{8}=\frac{1}{4}$

$\frac{1}{2}P=\frac{3}{8}$

$\frac{2}{1}\cdot\frac{1}{2}P=\frac{3}{8}\cdot\frac{2}{1}$

$P=\frac{6}{8}=\frac{3}{4}$

2. $\frac{1}{2}\cdot\frac{3}{4}-\frac{1}{8}=\frac{1}{4}$

$\frac{3}{8}-\frac{1}{8}=\frac{1}{4}$

$\frac{1}{4}=\frac{1}{4}$

3. $\frac{1}{3}B-\frac{2}{7}=1\frac{5}{7}$

$\frac{1}{3}B=2$

$\frac{3}{1}\cdot\frac{1}{3}B=2\cdot\frac{3}{1}$

$B=6$

4. $\frac{1}{3}(6)-\frac{2}{7}=1\frac{5}{7}$

$2-\frac{2}{7}=1\frac{5}{7}$

$1\frac{5}{7}=1\frac{5}{7}$

5. $\frac{3}{8}N+\frac{2}{8}=\frac{5}{8}$

$\frac{3}{8}N=\frac{3}{8}$

$\frac{8}{3}\cdot\frac{3}{8}N=\frac{3}{8}\cdot\frac{8}{3}$

$N=1$

6. $\frac{3}{8}(1)+\frac{1}{4}=\frac{5}{8}$

$\frac{3}{8}+\frac{2}{8}=\frac{5}{8}$

$\frac{5}{8}=\frac{5}{8}$

7. $\frac{4}{5}A+\frac{1}{9}=4\frac{1}{9}$

$\frac{4}{5}A=4$

$\frac{5}{4}\cdot\frac{4}{5}A=\frac{4}{1}\cdot\frac{5}{4}$

$A=\frac{20}{4}=5$

8. $\frac{4}{5}(5)+\frac{1}{9}=4\frac{1}{9}$

$4+\frac{1}{9}=4\frac{1}{9}$

$4\frac{1}{9}=4\frac{1}{9}$

9. $\frac{3}{4}R-\frac{4}{8}=\frac{1}{8}$

$\frac{3}{4}R=\frac{5}{8}$

$\frac{4}{3}\cdot\frac{3}{4}R=\frac{5}{8}\cdot\frac{4}{3}$

$R=\frac{20}{24}=\frac{5}{6}$

10. $\frac{3}{4}\cdot\frac{5}{6}-\frac{1}{2}=\frac{1}{8}$

$\frac{15}{24}-\frac{4}{8}=\frac{1}{8}$

$\frac{5}{8}-\frac{4}{8}=\frac{1}{8}$

$\frac{1}{8}=\frac{1}{8}$

11. $\frac{1}{6}D-\frac{1}{6}=8\frac{5}{6}$

$\frac{1}{6}D=9$

$\frac{6}{1}\cdot\frac{1}{6}D=9\cdot\frac{6}{1}$

$D=54$

12. $\frac{1}{6}(54)-\frac{1}{6}=8\frac{5}{6}$

$9-\frac{1}{6}=8\frac{5}{6}$

$8\frac{5}{6}=8\frac{5}{6}$

Lesson Practice 30C

1. $\frac{5}{6}T-\frac{1}{3}=\frac{1}{3}$

$\frac{5}{6}T=\frac{2}{3}$

$\frac{6}{5}\cdot\frac{5}{6}T=\frac{2}{3}\cdot\frac{6}{5}$

$T=\frac{12}{15}=\frac{4}{5}$

2. $\frac{5}{6}\cdot\frac{4}{5}-\frac{1}{3}=\frac{1}{3}$

$\frac{20}{30}-\frac{1}{3}=\frac{1}{3}$

$\frac{2}{3}-\frac{1}{3}=\frac{1}{3}$

$\frac{1}{3}=\frac{1}{3}$

3. $\frac{3}{4}F-\frac{3}{5}=17\frac{2}{5}$

$\frac{3}{4}F=18$

$\frac{4}{3}\cdot\frac{3}{4}F=18\cdot\frac{4}{3}$

$F=\frac{72}{3}=24$

4. $\frac{3}{4}(24)-\frac{3}{5}=17\frac{2}{5}$

$18-\frac{3}{5}=17\frac{2}{5}$

$17\frac{2}{5}=17\frac{2}{5}$

5. $\frac{2}{3}Q+\frac{1}{6}=\frac{1}{3}$

$\frac{2}{3}Q=\frac{1}{6}$

$\frac{3}{2}\cdot\frac{2}{3}Q=\frac{1}{6}\cdot\frac{3}{2}$

$Q=\frac{3}{12}=\frac{1}{4}$

6. $\frac{2}{3}\cdot\frac{1}{4}+\frac{1}{6}=\frac{1}{3}$

$\frac{1}{6}+\frac{1}{6}=\frac{1}{3}$

$\frac{1}{3}=\frac{1}{3}$

7. $\frac{3}{4}C+\frac{1}{3}=15\frac{1}{3}$

$\frac{3}{4}C=15$

$\frac{4}{3}\cdot\frac{3}{4}C=15\cdot\frac{4}{3}$

$C=\frac{60}{3}=20$

8. $\frac{3}{4}(20)+\frac{1}{3}=15\frac{1}{3}$

$15+\frac{1}{3}=15\frac{1}{3}$

$15\frac{1}{3}=15\frac{1}{3}$

9. $\frac{3}{5}S+\frac{7}{10}=\frac{9}{10}$

$\frac{3}{5}S=\frac{2}{10}$

$\frac{5}{3}\cdot\frac{3}{5}S=\frac{2}{10}\cdot\frac{5}{3}$

$S=\frac{10}{30}=\frac{1}{3}$

10. $\frac{3}{5} \cdot \frac{1}{3} + \frac{7}{10} = \frac{9}{10}$
$\frac{1}{5} + \frac{7}{10} = \frac{9}{10}$
$\frac{2}{10} + \frac{7}{10} = \frac{9}{10}$
$\frac{9}{10} = \frac{9}{10}$

11. $\frac{1}{6}H - \frac{1}{3} = 1\frac{2}{3}$
$\frac{1}{6}H = 2$
$\frac{6}{1} \cdot \frac{1}{6}H = 2 \cdot \frac{6}{1}$
$H = 12$

12. $\frac{1}{6}(12) - \frac{1}{3} = 1\frac{2}{3}$
$2 - \frac{1}{3} = 1\frac{2}{3}$
$1\frac{2}{3} = 1\frac{2}{3}$

5. $\frac{2}{5} = \frac{40}{100} = .40 = 40\%$
6. $\frac{3}{4} = \frac{75}{100} = .75 = 75\%$
7. $\frac{15}{24} < \frac{16}{24}$
8. $\frac{10}{50} = \frac{10}{50}$
9. $\frac{48}{64} > \frac{36}{64}$
10. done
11. 230
12. 525
13. 1,610
14. XLIX
15. CCCLII
16. DLXXXIII
17. MMDLV
18. $\frac{7}{2} \times \frac{5}{2} \times \frac{3}{2} = \frac{105}{8} = 13\frac{1}{8}$ cu. ft

Systematic Review 30D

1. $\frac{3}{5}G + \frac{1}{2} = 9\frac{1}{2}$
$\frac{3}{5}G = 9$
$\frac{5}{3} \cdot \frac{3}{5}G = 9 \cdot \frac{5}{3}$
$G = \frac{45}{3} = 15$

2. $\frac{3}{5}(15) + \frac{1}{2} = 9\frac{1}{2}$
$9 + \frac{1}{2} = 9\frac{1}{2}$
$9\frac{1}{2} = 9\frac{1}{2}$

3. $\frac{4}{7}U + \frac{2}{7} = \frac{6}{7}$
$\frac{4}{7}U = \frac{4}{7}$
$\frac{7}{4} \cdot \frac{4}{7}U = \frac{4}{7} \cdot \frac{7}{4}$
$U = 1$

4. $\frac{4}{7}(1) + \frac{2}{7} = \frac{6}{7}$
$\frac{6}{7} = \frac{6}{7}$

Systematic Review 30E

1. $\frac{1}{3}E + \frac{5}{8} = 6\frac{5}{8}$
$\frac{1}{3}E = 6$
$\frac{3}{1} \cdot \frac{1}{3}E = 6 \cdot \frac{3}{1}$
$E = 18$

2. $\frac{1}{3}(18) + \frac{5}{8} = 6\frac{5}{8}$
$6 + \frac{5}{8} = 6\frac{5}{8}$
$6\frac{5}{8} = 6\frac{5}{8}$

3. $\frac{1}{8}V - \frac{1}{8} = \frac{3}{4}$
$\frac{1}{8}V = \frac{7}{8}$
$\frac{8}{1} \cdot \frac{1}{8}V = \frac{7}{8} \cdot \frac{8}{1}$
$V = \frac{56}{8} = 7$

4. $\frac{1}{8}(7)-\frac{1}{8}=\frac{3}{4}$
$\frac{7}{8}-\frac{1}{8}=\frac{3}{4}$
$\frac{6}{8}=\frac{3}{4}$
$\frac{3}{4}=\frac{3}{4}$
5. $\frac{1}{4}=\frac{25}{100}=.25=25\%$
6. $\frac{4}{5}=\frac{80}{100}=.80=80\%$
7. $\frac{2}{3}=\frac{4}{6}=\frac{6}{9}=\frac{8}{12}$
8. $\frac{5}{8}=\frac{10}{16}=\frac{15}{24}=\frac{20}{32}$
9. 86
10. 152
11. 3,500
12. LXXIV
13. CCXI
14. MDXXII
15. 3 ft
16. 2 pt
17. 4 qt
18. 5x5 = 25
19. $4X+16=64$
$4X=48$
$\frac{1}{4}\cdot 4X=48\cdot\frac{1}{4}$
$X=12$
20. $5\frac{3}{8}+5\frac{3}{8}+5\frac{3}{8}+5\frac{3}{8}=$
$20\frac{12}{8}=21\frac{4}{8}=21\frac{1}{2}$ in

Systematic Review 30F

1. $\frac{5}{6}J+\frac{2}{5}=10\frac{2}{5}$
$\frac{5}{6}J=10$
$\frac{6}{5}\cdot\frac{5}{6}J=10\cdot\frac{6}{5}$
$J=\frac{60}{5}=12$
2. $\frac{5}{6}(12)+\frac{2}{5}=10\frac{2}{5}$
$\frac{5}{6}\cdot\frac{12}{1}+\frac{2}{5}=10\frac{2}{5}$
$\frac{60}{6}+\frac{2}{5}=10\frac{2}{5}$
$10\frac{2}{5}=10\frac{2}{5}$
3. $\frac{3}{5}=\frac{60}{100}=.60=60\%$
4. $\frac{1}{2}=\frac{50}{100}=.50=50\%$
5. $\frac{32}{15}\div\frac{4}{3}=\frac{\overset{8}{\cancel{32}}}{\underset{5}{\cancel{15}}}\times\frac{\cancel{3}}{\cancel{4}}=\frac{8}{5}=1\frac{3}{5}$
6. $\frac{12}{\cancel{5}}\times\frac{\cancel{5}}{\cancel{4}}\times\frac{\overset{2}{\cancel{8}}}{7}=\frac{24}{7}=3\frac{3}{7}$
7. $7\frac{5}{10}-3\frac{8}{10}=$
$6\frac{15}{10}-3\frac{8}{10}=3\frac{7}{10}$
8. $3\frac{15}{40}+1\frac{32}{40}=4\frac{47}{40}=5\frac{7}{40}$
9. 91
10. 410
11. 1,135
12. LVIII
13. CDXII
14. MMCCLX
15. 16 oz
16. 12 in
17. 5,280 ft
18. 2,000 lb
19. $\frac{22}{7}(35^2)=\frac{22}{\cancel{7}}\cdot\frac{\overset{175}{\cancel{1225}}}{1}=$
$\frac{3850}{1}=3,850$ sq. mi
20. $\frac{2}{1}\cdot\frac{22}{\cancel{7}}\cdot\frac{\overset{5}{\cancel{35}}}{1}=220$ mi

Appendix A1

1. done
2. $9+11=20$
   $20\div 2=10$
   10x6 = 60 sq. in
3. $6+10=16$
   $16\div 2=8$
   8x4 = 32 sq. ft
4. $6+12=18$
   $18\div 2=9$
   9x7 = 63 sq. ft
5. $3+5=8$
   $8\div 2=4$
   4x2 = 8 sq. ft
6. $8+10=18$
   $18\div 2=9$
   9x5 = 45 sq. in
7. $7+9=16$
   $16\div 2=8$
   8x5 = 40 sq. in
8. $1+3=4$
   $4\div 2=2$
   2x2 = 4 sq. mi

Appendix A2

1. $2+4=6$
   $6\div 2=3$
   3x3 = 9 sq. in
2. $3+17=20$
   $20\div 2=10$
   10x6 = 60 sq. ft
3. $5+9=14$
   $14\div 2=7$
   7x5 = 35 sq. ft
4. $4+10=14$
   $14\div 2=7$
   7x3 = 21 sq. in
5. $8+12=20$
   $20\div 2=10$
   10x11 = 110 sq. ft
6. $7+11=18$
   $18\div 2=9$
   9x4 = 36 sq. in
7. $2+6=8$
   $8\div 2=4$
   4x6 = 24 sq. ft
   24 plants
8. $5+7=12$
   $12\div 2=6$
   6x6 = 36 sq. in

Test Solutions

Test 1

1. $\frac{1}{2}$ of 12 is 6
2. $\frac{3}{4}$ of 16 is 12
3. $15 \div 3 = 5;\ 5 \text{x} 1 = 5$
4. $20 \div 5 = 4;\ 4 \text{x} 2 = 8$
5. $16 \div 8 = 2;\ 2 \text{x} 1 = 2$
6. $8 \div 4 = 2;\ 2 \text{x} 3 = 6$
7. $10 \div 2 = 5;\ 5 \text{x} 1 = 5$
8. $36 \div 6 = 6;\ 6 \text{x} 3 = 18$
9. $49 \div 7 = 7;\ 7 \text{x} 2 = 14$
10. $40 \div 5 = 8;\ 8 \text{x} 4 = 32$
11. $8 + 5 + 8 + 5 = 26$ in
12. $24 + 12 + 24 + 12 = 72$ ft
13. $55 + 32 + 55 + 32 = 174$ in
14. $18 \div 6 = 3;\ 3 \text{x} 1 = 3$ people
15. $\$14.58 + \$6.75 = \$21.33$
16. $\$25.00 - \$21.33 = \$3.67$
17. $14 \div 7 = 2;\ 2 \text{x} 2 = 4$ days
18. $12 + 15 + 12 + 15 = 54$ ft

Test 2

1. $\frac{3}{4}$; three-fourths
2. $\frac{1}{5}$; one-fifth
3. $\frac{3}{6}$; three-sixths
4. $\frac{4}{5}$; 5 sections; 4 shaded
5. $\frac{1}{3}$; 3 sections, 1 shaded
6. one – half; 2 sections, 1 shaded
7. $25 \div 5 = 5;\ 5 \text{x} 2 = 10$
8. $80 \div 10 = 8;\ 8 \text{x} 1 = 8$
9. $42 \div 7 = 6;\ 6 \text{x} 3 = 18$
10. $81 + 32 + 81 + 32 = 226$ yd
11. $28 + 28 + 28 + 28 = 112$ ft
12. $\frac{3}{5}$ of 35 = 21 people
13. $10 + 10 + 10 + 10 = 40$ ft
14. $\frac{1}{5}$ of 40 = 8 ft
15. $\$16.45 - \$5.30 = \$11.15$

Test 3

1. $\frac{1}{6} + \frac{4}{6} = \frac{5}{6}$ or five-sixths
2. $\frac{3}{4} - \frac{1}{4} = \frac{2}{4}$ or two-fourths
3. $\frac{3}{5} + \frac{2}{5} = \frac{5}{5}$
4. $\frac{4}{6} + \frac{1}{6} = \frac{5}{6}$
5. $\frac{2}{10} + \frac{6}{10} = \frac{8}{10}$
6. $\frac{5}{8} - \frac{3}{8} = \frac{2}{8}$
7. $\frac{8}{9} - \frac{2}{9} = \frac{6}{9}$
8. $\frac{2}{3} - \frac{1}{3} = \frac{1}{3}$
9. $24 \div 8 = 3;\ 3 \text{x} 1 = 3$
10. $10 \div 5 = 2;\ 2 \text{x} 4 = 8$
11. $20 \div 10 = 2;\ 2 \text{x} 4 = 8$
12. $54 + 59 + 98 = 211$ ft
13. $12 + 8 + 12 + 8 = 40$ yd
14. $16 + 16 + 16 + 16 = 64$ in
15. $\frac{1}{6} + \frac{2}{6} = \frac{3}{6}$ of a pie
16. $\frac{3}{3} - \frac{2}{3} = \frac{1}{3}$ of the job
17. $12 \div 4 = 3;\ 3 \text{x} 3 = 9$ months

18. $\frac{1}{6} + \frac{2}{6} = \frac{3}{6}$ of the donuts
$12 \div 6 = 2$
$2x3 = 6$
$\frac{3}{6}$ of $12 = 6$ donuts

19. $\frac{6}{6} - \frac{3}{6} = \frac{3}{6}$ of the donuts
$\frac{3}{6}$ of $12 = 6$ donuts

20. $\$15.50 + \$5.25 + \$10 = \$30.75$

Test 4

1. $\frac{3}{4} = \frac{6}{8} = \frac{9}{12} = \frac{12}{16} = \frac{15}{20}$
three-fourths = six-eighths = nine-twelths = twelve-sixteenths = fifteen-twentieth
2. $\frac{1}{5} = \frac{2}{10} = \frac{3}{15} = \frac{4}{20}$
3. $\frac{5}{6} = \frac{10}{12} = \frac{15}{18} = \frac{20}{24}$
4. $\frac{3}{7} = \frac{6}{14} = \frac{9}{21} = \frac{12}{28}$
5. $\frac{1}{2} = \frac{2}{4} = \frac{3}{6} = \frac{4}{8}$
6. $\frac{4}{5} - \frac{1}{5} = \frac{3}{5}$
7. $\frac{5}{8} + \frac{2}{8} = \frac{7}{8}$
8. $\frac{5}{9} - \frac{4}{9} = \frac{1}{9}$
9. $48 \div 8 = 6$; $6x3 = 18$
10. $18 \div 6 = 3$; $3x1 = 3$
11. $27 \div 3 = 9$; $9x2 = 18$
12. $22x31 = 682$
13. $54x11 = 594$
14. $322x22 = 7{,}084$
15. $\frac{1}{2} = \frac{2}{4}$ of the pages
16. $\frac{1}{2}$ of $20 = 10$ pages or
$\frac{2}{4}$ of $20 = 10$ pages
17. $13x12 = 156$ months
18. $11+11+11+11 = 44$ ft
19. $\frac{2}{7} + \frac{3}{7} = \frac{5}{7}$ read
$\frac{7}{7} - \frac{5}{7} = \frac{2}{7}$ left
20. $\frac{2}{7}$ of $56 = 16$
$56 \div 7 = 8$
$8x2 = 16$ pages

Test 5

1. $\frac{15}{30} + \frac{12}{30} = \frac{27}{30}$
2. $\frac{6}{12} + \frac{4}{12} = \frac{10}{12}$
3. $\frac{6}{30} + \frac{5}{30} = \frac{11}{30}$
4. $\frac{5}{20} - \frac{4}{20} = \frac{1}{20}$
5. $\frac{25}{30} - \frac{12}{30} = \frac{13}{30}$
6. $\frac{4}{8} - \frac{2}{8} = \frac{2}{8}$
7. $\frac{1}{7} = \frac{2}{14} = \frac{3}{21} = \frac{4}{28}$
8. $\frac{2}{3} = \frac{4}{6} = \frac{6}{9} = \frac{8}{12}$
9. $36 \div 9 = 4$; $4x5 = 20$
10. $90 \div 10 = 9$; $9x3 = 27$
11. $48 \div 6 = 8$; $8x4 = 32$
12. 90
13. 10
14. 300
15. 200
16. $\frac{3}{18} + \frac{6}{18} = \frac{9}{18}$ of the soldiers rode
17. $\frac{18}{18} - \frac{9}{18} = \frac{9}{18}$ of the soldiers walked
18. $\frac{4}{6} - \frac{3}{6} = \frac{1}{6}$ of a pound
19. $122x3 = 366$ cups
20. $7 + 24 + 25 = 56$ ft

Test 6

1. $\frac{5}{15}+\frac{6}{15}=\frac{11}{15}$
2. $\frac{18}{24}+\frac{4}{24}=\frac{22}{24}$
3. $\frac{30}{70}+\frac{7}{70}=\frac{37}{70}$
4. $\frac{6}{12}-\frac{2}{12}=\frac{4}{12}$
5. $\frac{15}{24}-\frac{8}{24}=\frac{7}{24}$
6. $\frac{20}{24}-\frac{18}{24}=\frac{2}{24}$
7. $\frac{7}{8}=\frac{14}{16}=\frac{21}{24}=\frac{28}{32}$
8. $\frac{1}{3}=\frac{2}{6}=\frac{3}{9}=\frac{4}{12}$
9. $50 \div 10 = 5$; $5 \times 1 = 5$
10. $16 \div 4 = 4$; $4 \times 3 = 12$
11. $20 \div 2 = 10$; $10 \times 1 = 10$
12. $(30) \times (30) = (900)$
    $26 \times 33 = 858$
13. $(80) \times (80) = (6{,}400)$
    $75 \times 83 = 6{,}225$
14. $(60) \times (20) = (1{,}200)$
    $61 \times 17 = 1{,}037$
15. $16 \times 12 = 192$ books
16. $\frac{1}{2}+\frac{1}{5}=\frac{5}{10}+\frac{2}{10}=\frac{7}{10}$ of his elephants
17. $\frac{4}{9}+\frac{3}{6}=\frac{24}{54}+\frac{27}{54}=\frac{51}{54}$ of the apples
18. $\frac{3}{4}-\frac{1}{8}=\frac{24}{32}-\frac{4}{32}=\frac{20}{32}$ of a pizza
19. 500
20. $\frac{7}{10} \times 60 = 42$ minutes

Test 7

1. $\frac{7}{28}<\frac{12}{28}$
2. $\frac{6}{16}<\frac{8}{16}$
3. $\frac{36}{45}>\frac{10}{45}$
4. $\frac{18}{33}<\frac{22}{33}$
5. $\frac{35}{63}<\frac{54}{63}$
6. $\frac{24}{32}=\frac{24}{32}$
7. $\frac{9}{72}+\frac{32}{72}=\frac{41}{72}$
8. $\frac{10}{15}-\frac{3}{15}=\frac{7}{15}$
9. $\frac{24}{30}+\frac{5}{30}=\frac{29}{30}$
10. $\frac{2}{5}=\frac{4}{10}=\frac{6}{15}=\frac{8}{20}$
11. $19 \div 2 = 9\frac{1}{2}$
12. $51 \div 6 = 8\frac{3}{6}$
13. $39 \div 5 = 7\frac{4}{5}$
14. $(40) \times (20) = (800)$
    $39 \times 24 = 936$
15. $(70) \times (20) = (1{,}400)$
    $72 \times 15 = 1{,}080$
16. $(70) \times (40) = (2{,}800)$
    $68 \times 43 = 2{,}924$
17. $\frac{5}{8}$ of 16 = 10 maple trees
18. $\frac{10}{50}<\frac{15}{50}$ Douglas ate more
19. $\frac{1}{5}$ of 20 = 4 Christa
    $\frac{3}{10}$ of 20 = 6 Douglas
    $4 < 6$; yes
20. $\frac{8}{10}-\frac{2}{10}=\frac{6}{10}$ inch more

Test 8

1. $\frac{7}{14}+\frac{6}{14}=\frac{13}{14}$
   $\frac{13}{14}+\frac{1}{3}=\frac{39}{42}+\frac{14}{42}=\frac{53}{42}=1\frac{11}{42}$
2. $\frac{16}{24}+\frac{6}{24}=\frac{22}{24}$
   $\frac{22}{24}+\frac{2}{3}=\frac{66}{72}+\frac{48}{72}=\frac{114}{72}=1\frac{42}{72}$
3. $\frac{6}{8}+\frac{3}{8}+\frac{4}{8}=\frac{13}{8}=1\frac{5}{8}$
4. $\frac{18}{24}+\frac{4}{24}=\frac{22}{24}$
   $\frac{22}{24}+\frac{1}{5}=\frac{110}{120}+\frac{24}{120}=\frac{134}{120}=1\frac{14}{120}$
5. $\frac{4}{16}+\frac{8}{16}=\frac{12}{16}$
   $\frac{12}{16}+\frac{2}{5}=\frac{60}{80}+\frac{32}{80}=\frac{92}{80}=1\frac{12}{80}$
6. $\frac{3}{6}+\frac{4}{6}+\frac{5}{6}=\frac{12}{6}=2$
7. $\frac{16}{32}=\frac{16}{32}$
8. $\frac{24}{56}>\frac{21}{56}$
9. $\frac{8}{20}<\frac{15}{20}$
10. $\frac{3}{9}=\frac{6}{18}=\frac{9}{27}=\frac{12}{36}$
11. $\frac{2}{3}=\frac{4}{6}=\frac{6}{9}=\frac{8}{12}$
12. $298\div 3=99\frac{1}{3}$
13. $156\div 7=22\frac{2}{7}$
14. $465\div 4=116\frac{1}{4}$
15. $\frac{2}{8}+\frac{4}{8}+\frac{1}{8}=\frac{7}{8}$ of his plants
16. $\frac{8}{8}-\frac{7}{8}=\frac{1}{8}$ alive; $\frac{1}{8}$ of 32 = 4 plants
17. 4x65 = 260 tomatoes
18. $217\div 5=\$43\frac{2}{5}$
19. 60
20. 800

Unit Test I

1. $12\div 3=4$; $4\text{x}2=8$
2. $10\div 5=2$; $2\text{x}3=6$
3. $24\div 4=6$; $6\text{x}3=18$
4. $15\div 3=5$; $5\text{x}1=5$
5. $25\div 5=5$; $5\text{x}4=20$
6. $14\div 7=2$; $2\text{x}5=10$
7. $\frac{2}{6}=\frac{4}{12}=\frac{6}{18}=\frac{8}{24}$
8. $\frac{5}{8}=\frac{10}{16}=\frac{15}{24}=\frac{20}{32}$
9. $\frac{1}{3}+\frac{1}{3}=\frac{2}{3}$
10. $\frac{6}{24}+\frac{12}{24}=\frac{18}{24}$
11. $\frac{6}{30}+\frac{25}{30}=\frac{31}{30}=1\frac{1}{30}$
    (final step optional)
12. $\frac{10}{15}-\frac{3}{15}=\frac{7}{15}$
13. $\frac{7}{10}-\frac{3}{10}=\frac{4}{10}$
14. $\frac{32}{56}-\frac{21}{56}=\frac{11}{56}$
15. $\frac{21}{35}<\frac{25}{35}$
16. $\frac{24}{48}=\frac{24}{48}$
17. $\frac{16}{24}>\frac{12}{24}$
18. $\frac{30}{80}+\frac{8}{80}=\frac{38}{80}$
    $\frac{38}{80}+\frac{2}{5}=\frac{190}{400}+\frac{160}{400}=\frac{350}{400}$
19. $\frac{6}{8}+\frac{7}{8}+\frac{4}{8}=\frac{17}{8}=2\frac{1}{8}$
    (final step optional)
20. 16 + 17 + 28 = 61 ft
21. 46 + 23 + 46 + 23 = 138 yd
22. 52 + 52 + 52 + 52 = 208 in
23. 90
24. 50
25. 600
26. 500
27. $358\div 5=71\frac{3}{5}$

28. $541 \div 8 = 67\frac{5}{8}$
29. $189 \div 6 = 31\frac{3}{6}$
30. $67 \times 18 = 1{,}206$
31. $34 \times 26 = 884$
32. $224 \times 22 = 4{,}928$
33. $\frac{2}{5} + \frac{1}{3} = \frac{6}{15} + \frac{5}{15} = \frac{11}{15}$ of her money
34. $\frac{15}{18} > \frac{12}{18}$ so $\frac{5}{6} > \frac{2}{3}$
    Drumore received more snow.
    $\frac{15}{18} - \frac{12}{18} = \frac{3}{18}$ ft difference
35. $10 + 11 + 10 + 11 = 42$ ft
    $\frac{1}{7}$ of $42 = 6$ ft of shelves

14. $(100) \times (40) = (4{,}000)$
    $124 \times 36 = 4{,}464$
15. $(1000) \times (10) = (10{,}000)$
    $957 \times 13 = 12{,}441$
16. $\frac{1}{3} + \frac{2}{5} = \frac{5}{15} + \frac{6}{15} = \frac{11}{15}$ of the customers
17. $10 + 12 + 12 = 34$ ft
18. $\frac{3}{8} \times \frac{1}{2} = \frac{3}{16}$ of the people
19. $\frac{3}{8}$ of 48:
    $48 \div 8 = 6$; $6 \times 3 = 18$ had burgers
    $\frac{3}{16}$ of 48:
    $48 \div 16 = 3$; $3 \times 3 = 9$ had mustard
20. $48 \times 15 = \$720$

Test 9

1. $\frac{1}{4}$ of $\frac{1}{3} = \frac{1}{12}$
2. $\frac{3}{4} \times \frac{2}{5} = \frac{6}{20}$
3. $\frac{2}{7} \times \frac{1}{4} = \frac{2}{28}$
4. $\frac{2}{5}$ of $\frac{3}{7} = \frac{6}{35}$
5. $\frac{3}{5} \times \frac{1}{6} = \frac{3}{30}$
6. $\frac{2}{4} \times \frac{1}{3} = \frac{2}{12}$
7. $\frac{12}{44} + \frac{11}{44} = \frac{23}{44}$
8. $\frac{24}{30} - \frac{5}{30} = \frac{19}{30}$
9. $\frac{3}{21} + \frac{14}{21} = \frac{17}{21}$
   $\frac{17}{21} + \frac{1}{2} = \frac{34}{42} + \frac{21}{42} = \frac{55}{42}$ or $1\frac{13}{42}$
10. $\frac{16}{24} > \frac{15}{24}$
11. $\frac{36}{48} = \frac{36}{48}$
12. $\frac{45}{54} > \frac{42}{54}$
13. $(600) \times (50) = (30{,}000)$
    $612 \times 54 = 33{,}048$

Test 10

1. $\frac{2}{5} \div \frac{1}{5} = \frac{2 \div 1}{1} = 2$
2. $\frac{15}{24} \div \frac{16}{24} = \frac{15 \div 16}{1} = \frac{15}{16}$
3. $\frac{4}{8} \div \frac{2}{8} = \frac{4 \div 2}{1} = 2$
4. $\frac{3}{12} \div \frac{4}{12} = \frac{3 \div 4}{1} = \frac{3}{4}$
5. $\frac{27}{45} \div \frac{10}{45} = \frac{27 \div 10}{1} = \frac{27}{10}$ or $2\frac{7}{10}$
6. $\frac{12}{15} \div \frac{10}{15} = \frac{12 \div 10}{1} = \frac{12}{10}$ or $1\frac{2}{10}$
7. $\frac{3}{4} \times \frac{1}{4} = \frac{3}{16}$
8. $\frac{2}{3} \times \frac{1}{5} = \frac{2}{15}$
9. $\frac{2}{9} \times \frac{1}{2} = \frac{2}{18}$
10. $\frac{32}{56} + \frac{21}{56} = \frac{53}{56}$
11. $\frac{20}{36} - \frac{9}{36} = \frac{11}{36}$
12. $\frac{4}{10} + \frac{3}{10} + \frac{5}{10} = \frac{12}{10}$ or $1\frac{2}{10}$
13. $(500) \div (40) \approx (10)$
14. $(900) \div (30) = (30)$
15. $(600) \div (60) = (10)$

16. $\frac{1}{2} \div \frac{1}{12} = \frac{12}{24} \div \frac{2}{24} = \frac{12 \div 2}{1} = 6$ pieces

17. $\frac{7}{8} \div \frac{1}{8} = \frac{7 \div 1}{1} = 7$ pieces

18. $\frac{3}{10} + \frac{1}{10} = \frac{4}{10}$

$\frac{4}{10} + \frac{3}{5} = \frac{20}{50} + \frac{30}{50} = \frac{50}{50} = 1$ mi

19. $\frac{7}{8} - \frac{2}{8} = \frac{5}{8}$ or $\frac{20}{32}$ of a pizza

20. $365 \text{x} 24 = 8{,}760$ hours

Test 11

1. no
2. yes
3. no
4. yes
5. 1, 2, 4, 8

   1, 2, 3, 4, 6, 8, 12, 24

   GCF = 8
6. 1, 2, 5, 10

   1, 2, 4, 5, 10, 20

   GCF = 10
7. 1, 3, 13, 39

   1, 3, 5, 15

   GCF = 3
8. 1, 2, 7, 14

   1, 2, 4, 7, 14, 28

   GCF = 14
9. $\frac{1}{8} \text{x} \frac{2}{7} = \frac{2}{56}$
10. $\frac{2}{9} \text{x} \frac{4}{5} = \frac{8}{45}$
11. $\frac{3}{6} \text{x} \frac{1}{6} = \frac{3}{36}$
12. $\frac{2}{16} \div \frac{8}{16} = \frac{2 \div 8}{1} = \frac{2}{8}$
13. $\frac{12}{18} \div \frac{3}{18} = \frac{12 \div 3}{1} = 4$
14. $\frac{50}{60} \div \frac{18}{60} = \frac{50 \div 18}{1} = \frac{50}{18}$ or $2\frac{14}{18}$
15. $\frac{36}{48} = \frac{36}{48}$
16. $\frac{30}{80} < \frac{32}{80}$
17. $\frac{9}{18} < \frac{10}{18}$
18. $673 \div 26 = 25\frac{23}{26}$
19. $390 \div 82 = 4\frac{62}{82}$
20. $768 \div 51 = 15\frac{3}{51}$

Test 12

1. $\frac{8}{10} \div \frac{2}{2} = \frac{4}{5}$
2. $\frac{12}{20} \div \frac{4}{4} = \frac{3}{5}$
3. $\frac{27}{30} \div \frac{3}{3} = \frac{9}{10}$
4. $\frac{20}{30} \div \frac{10}{10} = \frac{2}{3}$
5. $\frac{15}{25} \div \frac{5}{5} = \frac{3}{5}$
6. $\frac{36}{48} \div \frac{12}{12} = \frac{3}{4}$
7. yes
8. no
9. no
10. no
11. $\frac{12}{24} + \frac{6}{24} = \frac{18}{24} \div \frac{6}{6} = \frac{3}{4}$
12. $\frac{12}{54} + \frac{9}{54} = \frac{21}{54} \div \frac{3}{3} = \frac{7}{18}$
13. $\frac{2}{3} \text{x} \frac{5}{6} = \frac{10}{18} \div \frac{2}{2} = \frac{5}{9}$
14. $\frac{5}{8} \text{x} \frac{1}{10} = \frac{5}{80} \div \frac{5}{5} = \frac{1}{16}$
15. $\frac{10}{50} \div \frac{40}{50} = \frac{10 \div 40}{1} = \frac{10}{40} \div \frac{10}{10} = \frac{1}{4}$
16. $\frac{12}{21} \div \frac{14}{21} = \frac{12 \div 14}{1} = \frac{12}{14} \div \frac{2}{2} = \frac{6}{7}$
17. $573 \text{x} 612 = 350{,}676$
18. $3{,}682 \text{x} 694 = 2{,}555{,}308$
19. $\frac{3}{8} \text{x} \frac{1}{2} = \frac{3}{16}$ of the wall
20. $\frac{6}{8} \div \frac{1}{4} = \frac{24}{32} \div \frac{8}{32} = \frac{24 \div 8}{1} = 3$ parts

Test 13

1. 2x5x5
2. 2x2x2x2x3
3. 3x3x7
4. 3x3x11
5. $\frac{20}{30} = \frac{\cancel{2}\text{x}2\text{x}\cancel{5}}{\cancel{2}\text{x}3\text{x}\cancel{5}} = \frac{2}{3}$
6. $\frac{63}{81} = \frac{\cancel{3}\text{x}\cancel{3}\text{x}7}{\cancel{3}\text{x}\cancel{3}\text{x}3\text{x}3} = \frac{7}{9}$
7. $\frac{36}{42} = \frac{\cancel{2}\text{x}2\text{x}\cancel{3}\text{x}3}{\cancel{2}\text{x}\cancel{3}\text{x}7} = \frac{6}{7}$
8. $\frac{12}{72} + \frac{30}{72} = \frac{42}{72} \div \frac{6}{6} = \frac{7}{12}$
9. $\frac{30}{42} \div \frac{35}{42} = \frac{30 \div 35}{1} = \frac{30}{35} \div \frac{5}{5} = \frac{6}{7}$
10. $\frac{3}{4} \text{x} \frac{1}{3} = \frac{3}{12} \div \frac{3}{3} = \frac{1}{4}$
11. $\frac{32}{56} > \frac{21}{56}$
12. $\frac{55}{110} < \frac{60}{110}$
13. $\frac{15}{60} > \frac{8}{60}$
14. $5311 \div 83 = 63\frac{82}{83}$
15. $8856 \div 36 = 246$
16. $6532 \div 512 = 12\frac{388}{512}$ or $12\frac{97}{128}$
17. 25x17 = 425 qt
18. $\frac{1}{7} + \frac{7}{10} = \frac{10}{70} + \frac{49}{70} = \frac{59}{70}$ riding
    $\frac{70}{70} - \frac{59}{70} = \frac{11}{70}$ walking
19. $\frac{1}{2} \text{x} \frac{1}{3} = \frac{1}{6}$ bought backpacks
20. $\frac{12}{42} > \frac{7}{42}$ so $\frac{2}{7} > \frac{1}{6}$
    The group that bought dictionaries was larger.

Test 14

1. $\frac{4}{5}$
2. $\frac{4}{8} = \frac{1}{2}$
3. $\frac{10}{16} = \frac{5}{8}$
4. $\frac{2}{16} = \frac{1}{8}$
5. $\frac{13}{16}$
6. $\frac{5}{16}$
7. 2x2x2x3
8. 2x2x19
9. 2x2x2x2x3
10. $\frac{32}{54} = \frac{\cancel{2}\text{x}2\text{x}2\text{x}2\text{x}2}{\cancel{2}\text{x}3\text{x}3\text{x}3} = \frac{16}{27}$
11. 32x15 = 480 sq. in
12. 10x4 = 40 sq. ft
13. 49x24 = 1,176 sq. yd
14. no
15. 16 : $\underline{2}$, $\underline{4}$, 8 , 16
    44: $\underline{2}$, $\underline{4}$, 11 , 22, 44
    GCF = 4
16. $\frac{2}{5} \text{x} \frac{3}{4} = \frac{6}{20} \div \frac{2}{2} = \frac{3}{10}$ of the days
17. $650 \div 10 = 65$ bags
18. perimeter

Test 15

1. $\frac{9}{9} + \frac{2}{9} = \frac{11}{9}$
2. $\frac{5}{5} + \frac{5}{5} + \frac{5}{5} + \frac{3}{5} = \frac{18}{5}$
3. $\frac{3}{3} + \frac{1}{3} = \frac{4}{3}$
4. $\frac{4}{4} + \frac{4}{4} + \frac{1}{4} = \frac{9}{4}$
5. $\frac{4}{4} + \frac{3}{4} = 1\frac{3}{4}$
6. $\frac{2}{2} + \frac{2}{2} + \frac{2}{2} + \frac{2}{2} + \frac{1}{2} = 4\frac{1}{2}$
7. $\frac{5}{5} + \frac{5}{5} + \frac{3}{5} = 2\frac{3}{5}$

8. $\frac{6}{6}+\frac{6}{6}+\frac{6}{6}+\frac{5}{6}=3\frac{5}{6}$
9. $\frac{14}{16}=\frac{7}{8}$
10. $\frac{8}{16}=\frac{1}{2}$
11. $\frac{3}{9}\div\frac{3}{3}=\frac{1}{3}$
12. $\frac{20}{28}\div\frac{4}{4}=\frac{5}{7}$
13. $\frac{18}{32}\div\frac{2}{2}=\frac{9}{16}$
14. $16^2=256$ sq. in
15. $3^2=9$ sq. mi
16. $8^2=64$
17. $6^2=36$
18. yes
19. $15: \underline{3}, \underline{5}, \underline{15}$
    $30: 2, \underline{3}, \underline{5}, 6, 10, \underline{15}, 30$
    GCF = 15
20. 3x3x5

Test 16

1. $3\frac{3}{4}$
2. $2\frac{8}{10}=2\frac{4}{5}$
3. $\frac{8}{8}+\frac{1}{8}=\frac{9}{8}$
4. $\frac{6}{6}+\frac{6}{6}+\frac{5}{6}=\frac{17}{6}$
5. $\frac{4}{4}+\frac{4}{4}+\frac{1}{4}=2\frac{1}{4}$
6. $\frac{7}{7}+\frac{7}{7}+\frac{7}{7}+\frac{3}{7}=3\frac{3}{7}$
7. $\frac{27}{63}+\frac{14}{63}=\frac{41}{63}$
8. $\frac{6}{48}+\frac{8}{48}=\frac{14}{48}\div\frac{2}{2}=\frac{7}{24}$
9. $\frac{8}{24}-\frac{6}{24}=\frac{2}{24}\div\frac{2}{2}=\frac{1}{12}$
10. $\frac{5}{6}\times\frac{11}{12}=\frac{55}{72}$
11. $\frac{10}{20}\div\frac{6}{20}=\frac{10\div 6}{\not{20}\,1}=\frac{10}{6}=\frac{5}{3}=1\frac{2}{3}$
12. $\frac{2}{3}\times\frac{7}{20}=\frac{14}{60}\div\frac{2}{2}=\frac{7}{30}$
13. 10x4 = 40 sq ft
14. 10 + 4 + 10 + 4 = 28 ft
15. $23^2=529$ sq in
16. 23 + 23 + 23 + 23 = 92 in
17. $\frac{1}{2}(12)(5)=30$ sq ft
18. 5 + 12 + 13 = 30 ft
19. 16x16 = 256
20. $36: \underline{2}, \underline{3}, 4, \underline{6}, 9, 12, 18, 36$
    $42: \underline{2}, \underline{3}, \underline{6}, 7, 14, 21, 42$
    GCF = 6

Unit Test II

1. $\frac{9}{18}\div\frac{10}{18}=\frac{9\div 10}{1}=\frac{9}{10}$
2. $\frac{18}{42}\div\frac{35}{42}=\frac{18\div 35}{1}=\frac{18}{35}$
3. $\frac{56}{98}\div\frac{35}{98}=\frac{56\div 35}{1}=\frac{56}{35}=\frac{8}{5}=1\frac{3}{5}$
4. $\frac{21}{27}\div\frac{18}{27}=\frac{21\div 18}{1}=\frac{21}{18}=\frac{7}{6}=1\frac{1}{6}$
5. $\frac{24}{32}\div\frac{4}{32}=\frac{24\div 4}{1}=6$
6. $\frac{5}{20}\div\frac{12}{20}=\frac{5\div 12}{1}=\frac{5}{12}$
7. $\frac{1}{4}\times\frac{1}{3}=\frac{1}{12}$
8. $\frac{4}{5}\times\frac{7}{10}=\frac{28}{50}=\frac{14}{25}$
9. $\frac{2}{5}\times\frac{1}{2}=\frac{2}{10}=\frac{1}{5}$
10. $\frac{3}{4}\times\frac{1}{4}=\frac{3}{16}$
11. $\frac{3}{7}\times\frac{2}{9}=\frac{6}{63}=\frac{2}{21}$
12. $\frac{1}{8}\times\frac{1}{6}=\frac{1}{48}$
13. yes
14. no
15. yes
16. no
17. $28: \underline{2}, 4, 7, 14, 28$
    $54: \underline{2}, 3, 6, 9, 18, 27, 54$
    GCF = 2

18. 2x2x3x5
19. $\frac{4}{12} \div \frac{4}{4} = \frac{1}{3}$
20. $\frac{15}{20} \div \frac{5}{5} = \frac{3}{4}$
21. $\frac{18}{42} \div \frac{6}{6} = \frac{3}{7}$
22. $\frac{4}{4} + \frac{3}{4} = \frac{7}{4}$
23. $\frac{9}{9} + \frac{9}{9} + \frac{7}{9} = \frac{25}{9}$
24. $\frac{5}{5} + \frac{5}{5} + \frac{3}{5} = 2\frac{3}{5}$
25. $\frac{3}{3} + \frac{3}{3} + \frac{3}{3} + \frac{1}{3} = 3\frac{1}{3}$
26. $1\frac{10}{16} = 1\frac{5}{8}$
27. $4\frac{1}{10}$
28. $3\frac{2}{8} = 3\frac{1}{4}$
29. 7x3 = 21 sq. ft
30. $12^2 = 144$ sq. in
31. $\frac{1}{2}$ (8)(6) = 24 sq ft
32. 9x9 = 81
33. $\frac{3}{4} \div \frac{1}{16} = \frac{48}{64} \div \frac{4}{64} = \frac{48 \div 4}{1} = 12$ people
34. $\frac{3}{4} \times \frac{1}{2} = \frac{3}{8}$ of a cake

Test 17

1. $2\frac{1}{10} + 6\frac{6}{10} = 8\frac{7}{10}$
2. $4\frac{3}{5} - 3\frac{1}{5} = 1\frac{2}{5}$
3. $5\frac{2}{7} + 1\frac{4}{7} = 6\frac{6}{7}$
4. $6\frac{1}{2} = \frac{13}{2}$
5. $5\frac{2}{3} = \frac{17}{3}$
6. $1\frac{9}{10} = \frac{19}{10}$
7. $\frac{21}{4} = 5\frac{1}{4}$
8. $\frac{11}{5} = 2\frac{1}{5}$
9. $\frac{15}{8} = 1\frac{7}{8}$
10. $\frac{5}{6} \times \frac{1}{3} = \frac{5}{18}$
11. $\frac{1}{10} \div \frac{3}{10} = \frac{1 \div 3}{1} = \frac{1}{3}$
12. $\frac{8}{9} \times \frac{5}{6} = \frac{40}{54} = \frac{20}{27}$
13. 5x5x5 = 125 cu. in
14. 8x8x8 = 512 cu. ft
15. 13x13x13 = 2,197 cu. ft
16. 17x17x17 = 4,913 cu. ft
17. 5x19
18. $4\frac{3}{10} + 2\frac{6}{10} = 6\frac{9}{10}$ lb
19. $\frac{3}{4} \times \frac{3}{4} = \frac{9}{16}$ sq. mi
20. $\frac{5}{8}$ of 120

    120 ÷ 8 = 15; 15x5 = 75 parents

Test 18

1. $6\frac{4}{7} + 7\frac{5}{7}13\frac{9}{7} = 14\frac{2}{7}$
2. $13\frac{4}{9} + 5\frac{8}{9} = 18\frac{12}{9} = 19\frac{3}{9} = 19\frac{1}{3}$
3. $21\frac{11}{12} + 3\frac{1}{12} = 24\frac{12}{12} = 25$
4. $4\frac{3}{5} + 8\frac{4}{5} = 12\frac{7}{5} = 13\frac{2}{5}$
5. $9\frac{2}{3} + 8\frac{2}{3} = 17\frac{4}{3} = 18\frac{1}{3}$
6. $6\frac{7}{8} + 7\frac{3}{8} = 13\frac{10}{8} = 14\frac{2}{8} = 14\frac{1}{4}$
7. $\frac{3}{4} - \frac{1}{5} = \frac{15}{20} - \frac{4}{20} = \frac{11}{20}$
8. $\frac{1}{10} + \frac{1}{5} = \frac{5}{50} + \frac{10}{50} = \frac{15}{50} = \frac{3}{10}$
9. $\frac{7}{9} - \frac{1}{2} = \frac{14}{18} - \frac{9}{18} = \frac{5}{18}$
10. $\frac{3}{10} \div \frac{1}{10} = \frac{3 \div 1}{1} = 3$
11. $\frac{5}{6} \times \frac{1}{3} = \frac{5}{18}$

12. $\frac{3}{4} \div \frac{1}{3} = \frac{9}{12} \div \frac{4}{12}$

$= \frac{9 \div 4}{1} = \frac{9}{4} = 2\frac{1}{4}$

13. $64 \div 8 = 8;\ 8\text{x}3 = 24$
14. $42 \div 6 = 7;\ 7\text{x}1 = 7$
15. $80 \div 10 = 8;\ 8\text{x}7 = 56$
16. 8x5x2 = 80 cu. ft
17. 80x56 = 4,480 lb
18. 42: $\underline{2}$, 3, 6, $\underline{7}$, $\underline{14}$, 21, 42

    56: $\underline{2}$, 4, $\underline{7}$, 8, $\underline{14}$, 28, 56

    GCF = 14
19. 2x3x3
20. $2\frac{4}{5} + 1\frac{3}{5} = 3\frac{7}{5} = 4\frac{2}{5}$ mi

Test 19

1. $2\frac{1}{8} - 1\frac{5}{8} = 1\frac{9}{8} - 1\frac{5}{8} = \frac{4}{8} = \frac{1}{2}$
2. $5\frac{2}{7} - \frac{3}{7} = 4\frac{9}{7} - \frac{3}{7} = 4\frac{6}{7}$
3. $3 - 2\frac{1}{4} = 2\frac{4}{4} - 2\frac{1}{4} = \frac{3}{4}$
4. $2\frac{5}{8} + 2\frac{4}{8} = 4\frac{9}{8} = 5\frac{1}{8}$
5. $7\frac{3}{5} + 4\frac{4}{5} = 11\frac{7}{5} = 12\frac{2}{5}$
6. $8\frac{3}{6} + 9\frac{3}{6} = 17\frac{6}{6} = 18$
7. $\frac{4}{5} \div \frac{2}{3} = \frac{12}{15} \div \frac{10}{15} = \frac{12 \div 10}{1} =$

   $\frac{12}{10} = 1\frac{2}{10} = 1\frac{1}{5}$
8. $\frac{2}{11} \text{x} \frac{1}{5} = \frac{2}{55}$
9. $\frac{1}{2} + \frac{2}{3} + \frac{5}{6} = \frac{3}{6} + \frac{4}{6} + \frac{5}{6} = \frac{12}{6} = 2$
10. $\frac{5}{20} < \frac{8}{20}$
11. $\frac{24}{56} > \frac{7}{56}$
12. $\frac{50}{120} > \frac{48}{120}$
13. $99 \div 3 = 33$ yd
14. 26x3 = 78 ft
15. 9x3 = 27 ft
16. $\frac{2}{3} \text{x} \frac{1}{3} = \frac{2}{9}$ sq. mi
17. $\frac{2}{3} + \frac{2}{3} + \frac{1}{3} + \frac{1}{3} = \frac{6}{3} = 2$ mi
18. $9\frac{5}{8} + 6\frac{4}{8} = 15\frac{9}{8} = 16\frac{1}{8}$ pies
19. $14\frac{9}{8} - 11\frac{3}{8} = 3\frac{6}{8} = 3\frac{3}{4}$ yd
20. $6 \div 3 = 2$ yd

Test 20

1. $8\frac{5}{8} + \frac{1}{8} = 8\frac{6}{8}$

   $-3\frac{7}{8} + \frac{1}{8} = 4$

   $4\frac{6}{8} = 4\frac{3}{4}$
2. $14 + \frac{2}{3} = 14\frac{2}{3}$

   $-10\frac{1}{3} + \frac{2}{3} = 11$

   $3\frac{2}{3}$
3. $2\frac{3}{4} + 3\frac{3}{4} = 5\frac{6}{4} = 6\frac{2}{4} = 6\frac{1}{2}$
4. $7\frac{1}{3} - 3\frac{2}{3} = 6\frac{4}{3} - 3\frac{2}{3} = 3\frac{2}{3}$
5. $4\frac{3}{5} + 1\frac{4}{5} = 5\frac{7}{5} = 6\frac{2}{5}$
6. $\frac{2}{3} + \frac{3}{4} = \frac{8}{12} + \frac{9}{12} = \frac{17}{12} = 1\frac{5}{12}$
7. $\frac{3}{4} + \frac{5}{6} = \frac{18}{24} + \frac{20}{24}$

   $= \frac{38}{24} = \frac{19}{12} = 1\frac{7}{12}$
8. $\frac{1}{2} + \frac{4}{7} = \frac{7}{14} + \frac{8}{14} = \frac{15}{14} = 1\frac{1}{14}$
9. $\frac{7}{8} - \frac{1}{3} = \frac{21}{24} - \frac{8}{24} = \frac{13}{24}$
10. $\frac{5}{12} - \frac{2}{10} = \frac{50}{120} - \frac{24}{120}$

    $= \frac{26}{120} = \frac{13}{60}$
11. $\frac{1}{6} - \frac{1}{7} = \frac{7}{42} - \frac{6}{42} = \frac{1}{42}$
12. 32x2 = 64 pt
13. $34 \div 2 = 17$ qt

14. $7 \text{x} 3 = 21$ ft

15. $9\frac{3}{4} + 6\frac{3}{4} = 15\frac{6}{4}$
$= 16\frac{2}{4} = 16\frac{1}{2}$ tons

16. $4\frac{2}{5} - 1\frac{4}{5} = 3\frac{7}{5}$
$-1\frac{4}{5} = 2\frac{3}{5}$ gal

17. no

18. $9 \div 3 = 3$; $3 \text{x} 3 \text{x} 3 = 27$ cu. yd

19. $2\frac{1}{4} + 2\frac{1}{4} + 2\frac{1}{4} + 2\frac{1}{4} = 8\frac{4}{4} = 9$ ft

20. $15 \text{x} 15 = 225$

15. $54 \div 2 = 27$ qt

16. $18 \text{x} 2 = 36$ pt

17. $92 \div 3 = 30\frac{2}{3}$ yd

18. $20\frac{4}{8} + 2\frac{6}{8} = 22\frac{10}{8}$
$= 23\frac{2}{8} = 23\frac{1}{4}$ mi

19. $A = \frac{2}{3} \text{x} \frac{2}{3} = \frac{4}{9}$ sq. ft
$P = \frac{2}{3} + \frac{2}{3} + \frac{2}{3} + \frac{2}{3} = \frac{8}{3} = 2\frac{2}{3}$ ft

20. $3 \text{x} 4 = 12$
$12 \text{x} 2 = 24$
$24 + 1 = 25$
$25 \div 5 = 5$

Test 21

1. $9\frac{4}{12} + 6\frac{3}{12} = 15\frac{7}{12}$

2. $4\frac{2}{3} + 1\frac{3}{5} = 4\frac{10}{15} + 1\frac{9}{15} = 5\frac{19}{15} = 6\frac{4}{15}$

3. $9\frac{3}{5} + 2\frac{7}{10} = 9\frac{30}{50} + 2\frac{35}{50}$
$= 11\frac{65}{50} = 12\frac{15}{50} = 12\frac{3}{10}$

4. $12\frac{5}{11} + 4\frac{5}{8} = 12\frac{40}{88} + 4\frac{55}{88}$
$= 16\frac{95}{88} = 17\frac{7}{88}$

5. $5\frac{1}{5} - 2\frac{3}{5} = 4\frac{6}{5} - 2\frac{3}{5} = 2\frac{3}{5}$

6. $15\frac{7}{9} - 6\frac{2}{9} = 9\frac{5}{9}$

7. $7 - 3\frac{1}{5} = 6\frac{5}{5} - 3\frac{1}{5} = 3\frac{4}{5}$

8. $\frac{1}{2} \div \frac{1}{4} = \frac{4}{8} \div \frac{2}{8} = \frac{4 \div 2}{1} = 2$

9. $\frac{8}{12} \div \frac{1}{3} = \frac{24}{36} \div \frac{12}{36} = \frac{24 \div 12}{1} = 2$

10. $\frac{3}{5} \div \frac{1}{10} = \frac{30}{50} \div \frac{5}{50} = \frac{30 \div 5}{1} = 6$

11. $\frac{4}{6} \text{x} \frac{3}{5} = \frac{12}{30} = \frac{2}{5}$

12. $\frac{1}{2} \text{x} \frac{1}{3} = \frac{1}{6}$

13. $\frac{3}{4} \text{x} \frac{1}{6} = \frac{3}{24} = \frac{1}{8}$

14. $81 \div 3 = 27$ yd

Test 22

1. $4\frac{1}{4} - 2\frac{3}{4} = 3\frac{5}{4} - 2\frac{3}{4} = 1\frac{2}{4} = 1\frac{1}{2}$

2. $4\frac{1}{2} - 1\frac{1}{3} = 4\frac{3}{6} - 1\frac{2}{6} = 3\frac{1}{6}$

3. $8\frac{5}{9} - 3\frac{2}{9} = 5\frac{3}{9} = 5\frac{1}{3}$

4. $16\frac{3}{10} - 5\frac{2}{5} = 16\frac{15}{50} - 5\frac{20}{50} =$
$15\frac{65}{50} - 5\frac{20}{50} = 10\frac{45}{50} = 10\frac{9}{10}$

5. $1\frac{1}{2} + 3\frac{2}{3} = 1\frac{3}{6} + 3\frac{4}{6} = 4\frac{7}{6} = 5\frac{1}{6}$

6. $4\frac{2}{3} + 3\frac{4}{5} = 4\frac{10}{15} + 3\frac{12}{15}$
$= 7\frac{22}{15} = 8\frac{7}{15}$

7. $6\frac{3}{4} + 9\frac{7}{8} = 6\frac{24}{32} + 9\frac{28}{32} = 15\frac{52}{32}$
$= 15\frac{13}{8} = 16\frac{5}{8}$

8. $\frac{5}{16} \div \frac{3}{4} = \frac{20}{64} \div \frac{48}{64} = \frac{20 \div 48}{1}$
$= \frac{20}{48} = \frac{5}{12}$

9. $\frac{1}{2} \text{x} \frac{2}{3} = \frac{2}{6} = \frac{1}{3}$

10. $\frac{7}{8} \div \frac{5}{12} = \frac{84}{96} \div \frac{40}{96} = \frac{84 \div 40}{1}$
$= \frac{84}{40} = \frac{21}{10} = 2\frac{1}{10}$

11. $9\frac{5}{8} = \frac{77}{8}$
12. $25\frac{2}{3} = \frac{77}{3}$
13. $10\frac{3}{4} = \frac{43}{4}$
14. $14 \div 4 = 3\frac{1}{2}$ gal
15. $8x4 = 32$ qt
16. $20 \div 2 = 10$ qt
17. $672 \div 16 = 42$ yd
18. $16x3 = 48$ ft
19. $14: \underline{2}, \underline{7}, \underline{14}$
    $56: \underline{2}, 4, \underline{7}, 8, \underline{14}, 28, 56$
    GCF = 14
20. $9 + 1 = 10$
    $10 - 2 = 8$
    $8x7 = 56;\ 56 + 4 = 60$

Test 23

1. $\frac{3}{4} \div \frac{1}{2} = \frac{3}{4} x \frac{2}{1} = \frac{6}{4} = 1\frac{2}{4} = 1\frac{1}{2}$
2. $\frac{6}{8} \div \frac{4}{8} = \frac{6 \div 4}{1} = \frac{6}{4} = 1\frac{2}{4} = 1\frac{1}{2}$
3. $\frac{4}{5} \div \frac{2}{3} = \frac{4}{5} x \frac{3}{2} = \frac{12}{10} = 1\frac{2}{10} = 1\frac{1}{5}$
4. $\frac{12}{15} \div \frac{10}{15} = \frac{12 \div 10}{1} = \frac{12}{10} = 1\frac{2}{10} = 1\frac{1}{5}$
5. $\frac{5}{3} \div \frac{5}{11} = \frac{5}{3} x \frac{11}{5} = \frac{55}{15} = 3\frac{10}{15} = 3\frac{2}{3}$
6. $\frac{15}{8} \div \frac{1}{8} = \frac{15}{8} x \frac{8}{1} = \frac{120}{8} = 15$
7. $\frac{11}{4} \div \frac{1}{2} = \frac{11}{4} x \frac{2}{1} = \frac{22}{4} = 5\frac{2}{4} = 5\frac{1}{2}$
8. $\frac{31}{5} \div \frac{27}{8} = \frac{31}{5} x \frac{8}{27} = \frac{248}{135} = 1\frac{113}{135}$
9. $20\frac{4}{20} - 10\frac{15}{20} = 19\frac{24}{20} - 10\frac{15}{20} = 9\frac{9}{20}$
10. $9\frac{3}{24} - 4\frac{8}{24} = 8\frac{27}{24} - 4\frac{8}{24} = 4\frac{19}{24}$
11. $5\frac{12}{28} + 5\frac{7}{28} = 10\frac{19}{28}$
12. $17 \div 16 = \frac{17}{16} = 1\frac{1}{16}$ lb
13. $28 \div 2 = 14$ qt
14. $10x3 = 30$ ft
15. $5x16 = 80$ oz
16. $9x2 = 18$ pt
17. $19 \div 4 = \frac{19}{4} = 4\frac{3}{4}$ gal
18. $\frac{25}{8} \div \frac{5}{8} = \frac{25 \div 5}{1} = 5$ people
19. $4x2x1 = 8$ cu. ft
20. $6x7 = 42$
    $42 + 3 = 45$
    $45 \div 9 = 5$
    $5 + 2 = 7$

Unit Test III

1. $3\frac{3}{8} + 4\frac{1}{8} = 7\frac{4}{8} = 7\frac{1}{2}$
2. $7\frac{3}{5} + 4\frac{4}{5} = 11\frac{7}{5} = 12\frac{2}{5}$
3. $15\frac{4}{14} + 11\frac{7}{14} = 26\frac{11}{14}$
4. $8\frac{15}{18} + 5\frac{12}{18} = 13\frac{27}{18} =$
   $14\frac{9}{18} = 14\frac{1}{2}$
5. $9\frac{7}{8} - 3\frac{5}{8} = 6\frac{2}{8} = 6\frac{1}{4}$
6. $2\frac{5}{5} - 1\frac{2}{5} = 1\frac{3}{5}$
7. $35\frac{3}{12} - 21\frac{8}{12} =$
   $34\frac{15}{12} - 21\frac{8}{12} = 13\frac{7}{12}$
8. $7\frac{7}{42} - 6\frac{18}{42} =$
   $6\frac{49}{42} - 6\frac{18}{42} = \frac{31}{42}$
9. $48 \div 16 = 3$ lb
10. $23 \div 2 = 11\frac{1}{2}$ qt
11. $9x3 = 27$ ft
12. $4x16 = 64$ oz
13. $16x2 = 32$ pt
14. $20 \div 4 = 5$ gal
15. $\frac{3}{6} \div \frac{1}{2} = \frac{3}{6} x \frac{2}{1} = \frac{6}{6} = 1$
16. $\frac{6}{12} \div \frac{6}{12} = \frac{6 \div 6}{1} = 1$

17. $\frac{5}{8} \div \frac{2}{3} = \frac{5}{8} \times \frac{3}{2} = \frac{15}{16}$

18. $\frac{15}{24} \div \frac{16}{24} = \frac{15 \div 16}{1} = \frac{15}{16}$

19. $\frac{19}{3} \div \frac{19}{8} = \frac{19}{3} \times \frac{8}{19} = \frac{152}{57} = \frac{8}{3} = 2\frac{2}{3}$

20. $\frac{13}{5} \div \frac{19}{10} = \frac{13}{5} \times \frac{10}{19} = \frac{130}{95} = \frac{26}{19} = 1\frac{7}{19}$

21. $\frac{35}{6} \div \frac{5}{6} = \frac{35}{6} \times \frac{6}{5} = \frac{210}{30} = 7$

22. $\frac{7}{2} \div \frac{10}{7} = \frac{7}{2} \times \frac{7}{10} = \frac{49}{20} = 2\frac{9}{20}$

23. $\frac{1}{8}$

24. $15 \times 10 \times 6 = 900$ cu. ft

25. $\frac{9}{2} \div \frac{8}{1} = \frac{9}{2} \times \frac{1}{8} = \frac{9}{16}$ of a pizza

26. $2\frac{6}{18} + 1\frac{15}{18} = 3\frac{21}{18} = 4\frac{3}{18} = 4\frac{1}{6}$ mi

27. $5\frac{3}{6} - 2\frac{4}{6} = 4\frac{9}{6} - 2\frac{4}{6} = 2\frac{5}{6}$ rows

28. $8 \times 6 = 48$; $48 - 4 = 44$;
$44 \div 2 = 22$; $22 + 3 = 25$

10. $\frac{25}{6} \div \frac{24}{5} = \frac{25}{6} \times \frac{5}{24} = \frac{125}{144}$

11. $2\frac{4}{3} - 1\frac{2}{3} = 1\frac{2}{3}$

12. $8\frac{8}{12} + 5\frac{9}{12} = 13\frac{17}{12} = 14\frac{5}{12}$

13. $4\frac{2}{12} - 1\frac{6}{12} = 3\frac{14}{12} - 1\frac{6}{12}$
$= 2\frac{8}{12} = 2\frac{2}{3}$

14. $\frac{2}{3} \times \frac{3}{5} = \frac{6}{15} = \frac{2}{5}$

15. $\frac{2}{7} \times \frac{3}{4} = \frac{6}{28} = \frac{3}{14}$

16. $\frac{5}{8} \times \frac{1}{3} = \frac{5}{24}$

17. $6 \times 12 = 72$ in

18. $3J = 120$; $J = 40$ years

19. $\frac{1}{2} \times 12 = 6$ in

20. $A = \frac{2}{3} \times \frac{2}{3} = \frac{4}{9}$ sq. mi
$P = \frac{2}{3} + \frac{2}{3} + \frac{2}{3} + \frac{2}{3} = \frac{8}{3} = 2\frac{2}{3}$ mi

Test 24

1. $\frac{1}{10} \cdot 10J = 200 \cdot \frac{1}{10}$
$J = 20$

2. $10(20) = 200$
$200 = 200$

3. $\frac{1}{6} \cdot 6U = 24 \cdot \frac{1}{6}$
$U = 4$

4. $6(4) = 24$
$24 = 24$

5. $\frac{1}{5} \cdot 5K = 95 \cdot \frac{1}{5}$
$K = 19$

6. $5(19) = 95$
$95 = 95$

7. $\frac{1}{3} \div \frac{2}{5} = \frac{1}{3} \times \frac{5}{2} = \frac{5}{6}$

8. $\frac{5}{15} \div \frac{6}{15} = \frac{5 \div 6}{1} = \frac{5}{6}$

9. $\frac{5}{4} \div \frac{7}{2} = \frac{5}{4} \times \frac{2}{7} = \frac{10}{28} = \frac{5}{14}$

Test 25

1. $\frac{\cancel{2}}{\cancel{3}} \times \frac{\cancel{3}}{\cancel{4}_{2}} \times \frac{\cancel{2}}{7} = \frac{1}{7}$

2. $\frac{\overset{3}{\cancel{12}}}{\cancel{7}} \times \frac{1}{5} \times \frac{\cancel{7}}{\cancel{4}} = \frac{3}{5}$

3. $\frac{1}{\cancel{8}} \times \frac{\cancel{8}}{7} \times \frac{4}{7} = \frac{4}{49}$

4. $\frac{\cancel{6}}{\cancel{5}} \times \frac{\cancel{5}}{\cancel{6}} \times \frac{3}{2} = \frac{3}{2} = 1\frac{1}{2}$

5. $\frac{1}{7} \cdot 7L = 77 \cdot \frac{1}{7}$
$L = 11$

6. $7(11) = 77$
$77 = 77$

7. $\frac{1}{13} \cdot 13Y = 195 \cdot \frac{1}{13}$
$Y = 15$

8. $13(15) = 195$
$195 = 195$

9. $\frac{25}{16} \div \frac{5}{8} = \frac{\overset{5}{\cancel{25}}}{\underset{2}{\cancel{16}}} \times \frac{\cancel{8}}{\cancel{5}} = \frac{5}{2} = 2\frac{1}{2}$

10. $\frac{19}{3} \div \frac{19}{8} = \frac{\cancel{19}}{3} \times \frac{8}{\cancel{19}} = \frac{8}{3} = 2\frac{2}{3}$

11. $5\frac{13}{8} - 3\frac{7}{8} = 2\frac{6}{8} = 2\frac{3}{4}$

12. $11\frac{5}{4} - 11\frac{3}{4} = \frac{2}{4} = \frac{1}{2}$

13. $1\frac{1}{3} + 1\frac{2}{3} = 2\frac{3}{3} = 3$

14. $2 \text{x} 2{,}000 = 4{,}000$ lb

15. $24 \div 12 = 2$ ft

16. $\frac{1}{4}$ of $2{,}000 = 500$ lb

17. $\frac{3}{2} \text{x} \frac{2000}{1} = \frac{6000}{2} = 3{,}000$ lb

18. $\frac{\cancel{5}}{\underset{2}{\cancel{8}}} \text{x} \frac{1}{10} \text{x} \frac{\cancel{4}}{\cancel{5}} = \frac{1}{20}$ of the money

19. $\frac{13}{2} \div \frac{13}{1} = \frac{\cancel{13}}{2} \text{x} \frac{1}{\cancel{13}} = \frac{1}{2}$ dozen

$\frac{1}{2}$ of $12 = 6$ donuts per person

20. $28 \div 4 = 7$

$7 + 1 = 8$

$8 \text{x} 8 = 64$

$64 - 4 = 60$

Test 26

1. $6B + 10 = 40$

$6B = 30$

$\frac{1}{6} \cdot 6B = 30 \cdot \frac{1}{6}$

$B = 5$

2. $6(5) + 10 = 40$

$30 + 10 = 40$

$40 = 40$

3. $8N - 13 = 3$

$8N = 16$

$\frac{1}{8} \cdot 8N = 16 \cdot \frac{1}{8}$

$N = 2$

4. $8(2) - 13 = 3$

$16 - 13 = 3$

$3 = 3$

5. $6C + 8 = 50$

$6C = 42$

$\frac{1}{6} \cdot 6C = 42 \cdot \frac{1}{6}$

$C = 7$

6. $6(7) + 8 = 50$

$42 + 8 = 50$

$50 = 50$

7. $\frac{6}{\cancel{5}} \text{x} \frac{\overset{2}{\cancel{10}}}{\cancel{3}} \text{x} \frac{\cancel{3}}{11} = \frac{12}{11} = 1\frac{1}{11}$

8. $\frac{1}{\cancel{2}} \text{x} \frac{5}{\cancel{12}} \text{x} \frac{\overset{\cancel{12}}{\cancel{24}}}{7} = \frac{5}{7}$

9. $\frac{1}{5} \div \frac{8}{25} = \frac{1}{\cancel{5}} \text{x} \frac{\overset{5}{\cancel{25}}}{8} = \frac{5}{8}$

10. $\frac{20}{3} \div \frac{20}{7} = \frac{\cancel{20}}{3} \text{x} \frac{7}{\cancel{20}} = \frac{7}{3} = 2\frac{1}{3}$

11. $3 \text{ x } 5{,}280 = 15{,}840$ ft

12. $7 \text{x} 16 = 112$ oz

13. $\frac{15}{4} \text{x} 12 = \frac{180}{4} = 45$ in

14. $2\frac{1}{4} + 3\frac{1}{4} = 5\frac{2}{4} = 5\frac{1}{2}$ tons

15. no

16. 2x2x2x2x3

17. 18: $\underline{2}$, $\underline{3}$, $\underline{6}$, 9, 18

24: $\underline{2}$, $\underline{3}$, 4, $\underline{6}$, 8, 12, 24

GCF = 6

18. $\frac{3}{2} \text{x} \frac{3}{2} = \frac{9}{4} = 2\frac{1}{4}$ sq. mi

19. $16 \text{x} 16 = 256$

20. $56 \div 7 = 8$

$8 + 10 = 18$

$18 - 3 = 15$

$15 + 15 = 30$

Test 27

1. $\frac{22}{7}\left(35^2\right) = \frac{22}{\cancel{7}} \times \frac{\overset{175}{\cancel{1225}}}{1} =$

   $\frac{3850}{1} = 3,850$ sq in
2. $\frac{2}{1} \times \frac{22}{\cancel{7}} \times \frac{\overset{5}{\cancel{35}}}{1} = \frac{220}{1} = 220$ in
3. $\frac{22}{7}\left(5^2\right) = \frac{22}{7} \times \frac{25}{1} = \frac{550}{7}$

   $= 78\frac{4}{7}$ sq ft
4. $\frac{2}{1} \times \frac{22}{7} \times \frac{5}{1} = \frac{220}{7} = 31\frac{3}{7}$ ft
5. $\frac{22}{7}\left(63^2\right) = \frac{22}{\cancel{7}} \times \frac{\overset{567}{\cancel{3969}}}{1} =$

   $\frac{12474}{1} = 12,474$ sq in
6. $\frac{2}{1} \times \frac{22}{\cancel{7}} \times \frac{\overset{9}{\cancel{63}}}{1} = \frac{396}{1} = 396$ in
7. $\frac{22}{7}\left(3^2\right) = \frac{22}{7} \times \frac{9}{1} = \frac{198}{7}$

   $= 28\frac{2}{7}$ sq ft
8. $\frac{2}{1} \times \frac{22}{7} \times \frac{3}{1} = \frac{132}{7} = 18\frac{6}{7}$ ft
9. $5R - 17 = 38$

   $5R = 55$

   $\frac{1}{5} \cdot 5R = 55 \cdot \frac{1}{5}$

   $R = 11$
10. $5(11) - 17 = 38$

    $55 - 17 = 38$

    $38 = 38$
11. $\frac{\overset{2}{\cancel{4}}}{\cancel{3}} \times \frac{\cancel{3}}{\cancel{2}} = \frac{2}{1} = 2$
12. $\frac{\cancel{5}}{\cancel{3}} \times \frac{7}{\cancel{3}} \times \frac{\overset{3}{\cancel{9}}}{\underset{2}{\cancel{10}}} = \frac{7}{2} = 3\frac{1}{2}$
13. $\frac{7}{10} \div \frac{7}{12} = \frac{\cancel{7}}{\underset{5}{\cancel{10}}} \times \frac{\overset{6}{\cancel{12}}}{\cancel{7}}$

    $= \frac{6}{5} = 1\frac{1}{5}$
14. $\frac{22}{5} \div \frac{7}{9} = \frac{22}{5} \times \frac{9}{7} = \frac{198}{35} = 5\frac{23}{35}$
15. 19 qt < 20 qt
16. 6,000 lb > 5,000 lb
17. 3,520 ft < 3,720 ft
18. $8\frac{4}{6} - 6\frac{3}{6} = 2\frac{1}{6}$ mi
19. $\frac{13}{6} \times 5,280 = \frac{68,640}{6} = 11,440$ ft
20. $\frac{1}{\cancel{2}} \times \frac{\overset{\cancel{2}}{\cancel{4}}}{1} \times \frac{5}{\cancel{2}} = \frac{5}{1} = 5$ sq. in

Test 28

1. $\frac{4}{15}W + 7 = 19$

   $\frac{4}{15}W = 12$

   $\frac{15}{4} \cdot \frac{4}{15}W = \frac{12}{1} \cdot \frac{15}{4}$

   $W = \frac{180}{4}$

   $W = 45$
2. $\frac{4}{15}(45) + 7 = 19$

   $\frac{4}{15} \cdot \frac{45}{1} + 7 = 19$

   $\frac{180}{15} + 7 = 19$

   $12 + 7 = 19$

   $19 = 19$
3. $\frac{2}{3}R - 5 = 5$

   $\frac{2}{3}R = 10$

   $\frac{3}{2} \cdot \frac{2}{3}R = \frac{10}{1} \cdot \frac{3}{2}$

   $R = \frac{30}{2} = 15$
4. $\frac{2}{3}(15) - 5 = 5$

   $\frac{2}{3} \cdot \frac{15}{1} - 5 = 5$

   $\frac{30}{3} - 5 = 5$

   $10 - 5 = 5$

   $5 = 5$

5. $\frac{2}{1}\cdot\frac{1}{2}X=\frac{4}{1}\cdot\frac{2}{1}$

$X=\frac{8}{1}=8$

6. $\frac{1}{2}\cdot\frac{8}{1}=4$

$\frac{8}{2}=4$

$4=4$

7. $\frac{\overset{5}{\cancel{15}}}{\cancel{8}}\times\frac{\cancel{8}}{\cancel{3}}=\frac{5}{1}=5$

8. $\frac{\cancel{4}}{\cancel{5}}\times\frac{\overset{5}{\cancel{10}}}{\cancel{3}}\times\frac{\overset{\cancel{3}}{\cancel{15}}}{\underset{\cancel{2}}{\cancel{8}}}=\frac{5}{1}=5$

9. $\frac{11}{18}\div\frac{3}{8}=\frac{11}{\underset{9}{\cancel{18}}}\times\frac{\overset{4}{\cancel{8}}}{3}=\frac{44}{27}=1\frac{17}{27}$

10. $\frac{9}{2}\div\frac{5}{2}=\frac{9\div5}{1}=\frac{9}{5}=1\frac{4}{5}$

11. 48 oz > 32 oz

12. 1,056 ft = 1,056 ft

13. 1 qt < 2 qt

14. $2+4+6+8=20$

$20\div4=5$

15. $1+3+5+7=16$

$16\div4=4$

16. $8+10+12=30$

$30\div3=10$

17. $\frac{22}{7}(28^2)=\frac{22}{7}\cdot\frac{784}{1}=$

$\frac{2464}{1}=2{,}464$ sq. yd

18. $\frac{2}{1}\times\frac{22}{\cancel{7}}\times\frac{\overset{4}{\cancel{28}}}{1}=\frac{176}{1}=176$ yd

19. $\frac{11}{4}\times\frac{11}{4}\times\frac{11}{4}=\frac{1331}{64}=$

$20\frac{51}{64}$ cu. in

20. $9\times8=72$

$72-2=70$

$70\div10=7$

$7\times4=28$

Test 29

1. $\frac{5}{5}=\frac{100}{100}=100\%$

2. $\frac{3}{5}=\frac{60}{100}=60\%$

3. $\frac{4}{5}=\frac{80}{100}=.80=80\%$

4. $\frac{3}{4}=\frac{75}{100}=.75=75\%$

5. $\frac{3}{5}Y+2=26$

$\frac{3}{5}Y=24$

$\frac{5}{3}\cdot\frac{3}{5}Y=\frac{24}{1}\cdot\frac{5}{3}$

$Y=\frac{120}{3}=40$

6. $\frac{3}{5}(40)+2=26$

$\frac{3}{5}\cdot\frac{40}{1}+2=26$

$\frac{120}{5}+2=26$

$24+2=26$

$26=26$

7. $6\frac{7}{5}-2\frac{4}{5}=4\frac{3}{5}$

8. $8\frac{4}{4}-6\frac{3}{4}=2\frac{1}{4}$

9. $10\frac{72}{80}+4\frac{30}{80}=14\frac{102}{80}=$

$15\frac{22}{80}=15\frac{11}{40}$

10. 10

11. 14

12. 22

13. 35

14. III

15. XVII

16. XXV

17. XIX

18. $15+10+8+25=\$58$

$\$58\div4=14\frac{2}{4}=\$14\frac{1}{2}$ per day

19. $A = \frac{22}{7}(14^2) = \frac{22}{\cancel{7}} \cdot \frac{\overset{28}{\cancel{196}}}{1} =$

$\frac{616}{1} = 616$ sq. in

$C = \frac{2}{1} \cdot \frac{22}{\cancel{7}} \cdot \frac{\overset{2}{\cancel{14}}}{1} = \frac{88}{1} = 88$ in

20. $\frac{1}{2} \cdot \frac{16}{1} \cdot \frac{23}{4} = \frac{368}{8} = 46$ sq. ft

Test 30

1. $\frac{4}{9}W + \frac{1}{3} = \frac{2}{3}$

   $\frac{4}{9}W = \frac{1}{3}$

   $\frac{9}{4} \cdot \frac{4}{9}W = \frac{1}{3} \cdot \frac{9}{4}$

   $W = \frac{9}{12} = \frac{3}{4}$

2. $\frac{4}{9} \cdot \frac{3}{4} + \frac{1}{3} = \frac{2}{3}$

   $\frac{12}{36} + \frac{1}{3} = \frac{2}{3}$

   $\frac{1}{3} + \frac{1}{3} = \frac{2}{3}$

   $\frac{2}{3} = \frac{2}{3}$

3. $\frac{1}{2}K - \frac{3}{4} = 4\frac{1}{4}$

   $\frac{1}{2}K = 5$

   $\frac{2}{1} \cdot \frac{1}{2}K = 5 \cdot \frac{2}{1}$

   $K = 10$

4. $\frac{1}{2}(10) - \frac{3}{4} = 4\frac{1}{4}$

   $5 - \frac{3}{4} = 4\frac{1}{4}$

   $4\frac{1}{4} = 4\frac{1}{4}$

5. $\frac{1}{5} = \frac{20}{100} = .20 = 20\%$

6. $\frac{1}{4} = \frac{25}{100} = .25 = 25\%$

7. $\frac{15}{8} \div \frac{5}{8} = \frac{15 \div 5}{1} = 3$

8. $\frac{\overset{2}{\cancel{10}}}{\cancel{8}} \times \frac{\overset{2}{\cancel{16}}}{\cancel{5}} \times \frac{26}{5} = \frac{104}{5} = 20\frac{4}{5}$

9. $8\frac{12}{18} - 4\frac{15}{18} = 7\frac{30}{18} - 4\frac{15}{18} =$

   $3\frac{15}{18} = 3\frac{5}{6}$

10. $11\frac{9}{18} + 6\frac{10}{18} = 17\frac{19}{18} = 18\frac{1}{18}$

11. 51

12. 115

13. 1,600

14. XI

15. CCC

16. MDCC

17. 5,280 x 2 = 10,560 ft

18. 2,000 x 3 = 6,000 lb

19. $\frac{15}{1} \times \frac{\overset{5}{\cancel{10}}}{1} \times \frac{7}{\cancel{2}} = 525$ cu. ft

20. $8X - 8 = 40$

    $8X = 48$

    $\frac{1}{8} \cdot 8X = 48 \cdot \frac{1}{8}$

    $X = 6$

Unit Test IV

1. $7L = 49$

   $\frac{1}{7} \cdot 7L = 49 \cdot \frac{1}{7}$

   $L = 7$

2. $7(7) = 49$

   $49 = 49$

3. $3R - 8 = 22$

   $3R = 30$

   $\frac{1}{3} \cdot 3R = 30 \cdot \frac{1}{3}$

   $R = 10$

4. $3(10) - 8 = 22$

   $30 - 8 = 22$

   $22 = 22$

5. $\frac{5}{8}X+9=19$

$\frac{5}{8}X=10$

$\frac{8}{5}\cdot\frac{5}{8}X=10\cdot\frac{8}{5}$

$X=\frac{80}{5}=16$

6. $\frac{5}{8}(16)+9=19$

$\frac{5}{8}\cdot\frac{16}{1}+9=19$

$\frac{80}{8}+9=19$

$10+9=19$

$19=19$

7. $\frac{3}{10}Y-\frac{1}{5}=\frac{2}{5}$

$\frac{3}{10}Y=\frac{3}{5}$

$\frac{10}{3}\cdot\frac{3}{10}Y=\frac{3}{5}\cdot\frac{10}{3}$

$Y=\frac{30}{15}=2$

8. $\frac{3}{10}(2)-\frac{1}{5}=\frac{2}{5}$

$\frac{3}{10}\cdot\frac{2}{1}-\frac{1}{5}=\frac{2}{5}$

$\frac{6}{10}-\frac{1}{5}=\frac{2}{5}$

$\frac{3}{5}-\frac{1}{5}=\frac{2}{5}$

$\frac{2}{5}=\frac{2}{5}$

9. $\frac{3}{4}T+\frac{3}{8}=6\frac{3}{8}$

$\frac{3}{4}T=6$

$\frac{4}{3}\cdot\frac{3}{4}T=6\cdot\frac{4}{3}$

$T=\frac{24}{3}=8$

10. $\frac{3}{4}\cdot\frac{8}{1}+\frac{3}{8}=6\frac{3}{8}$

$\frac{24}{4}+\frac{3}{8}=6\frac{3}{8}$

$6+\frac{3}{8}=6\frac{3}{8}$

$6\frac{3}{8}=6\frac{3}{8}$

11. $\frac{\overset{3}{\cancel{9}}}{\underset{2}{\cancel{10}}}\times\frac{\cancel{5}}{7}\times\frac{\cancel{2}}{\cancel{3}}=\frac{3}{7}$

12. $\frac{19}{\cancel{9}}\times\frac{\cancel{9}}{\cancel{4}}\times\frac{\overset{3}{\cancel{12}}}{11}=$

$\frac{57}{11}=5\frac{2}{11}$

13. $\frac{3}{4}=\frac{75}{100}=.75=75\%$

14. $\frac{1}{2}=\frac{50}{100}=.50=50\%$

15. $\frac{22}{7}\left(49^2\right)=\frac{22}{7}\cdot\frac{2401}{1}=$

$\frac{7546}{1}=7{,}546$ sq. in

16. $\frac{2}{1}\cdot\frac{22}{\cancel{7}}\cdot\frac{\overset{7}{\cancel{49}}}{1}=\frac{308}{1}=308$ in

17. 10
18. 1
19. 50
20. 1,000
21. 500
22. 5
23. 100
24. 5,280 x 5 = 26,400 ft

25. $\frac{13}{\cancel{2}}\times\frac{\overset{1000}{\cancel{2000}}}{1}=13{,}000$ lb

26. $\frac{23}{\cancel{4}}\times\frac{\overset{3}{\cancel{12}}}{1}=69$ in

27. $169\div12=14\frac{1}{12}$ ft

28. $100+90+85+97=372$

$372\div4=93$ is average score

Final Test

1. $24\div2=12$

$12\times1=12$

2. $18\div3=6$

$6\times2=12$

3. $64\div8=8$

$8\times7=56$

4. $\frac{3}{4} = \frac{6}{8} = \frac{9}{12} = \frac{12}{16}$

5. $\frac{9}{10} = \frac{18}{20} = \frac{27}{30} = \frac{36}{40}$

6. $\frac{25}{35} > \frac{21}{35}$

7. $\frac{24}{48} = \frac{24}{48}$

8. $\frac{12}{24} < \frac{16}{24}$

9. $\frac{3}{9} + \frac{5}{9} = \frac{8}{9}$

10. $\frac{4}{8} + \frac{2}{8} + \frac{7}{8} = \frac{13}{8} = 1\frac{5}{8}$

11. $\frac{12}{15} - \frac{5}{15} = \frac{7}{15}$

12. $\frac{1}{3} \times \frac{5}{1} = \frac{5}{3} = 1\frac{2}{3}$

13. $\frac{\cancel{10}^{\,2}}{\cancel{3}} \times \frac{\cancel{18}^{\,6}}{\cancel{5}} = 12$

14. $\frac{\cancel{19}}{\cancel{5}} \times \frac{\cancel{25}^{\,5}}{\cancel{57}_{\,3}} = \frac{5}{3} = 1\frac{2}{3}$

15. $7\frac{1}{4} - 5\frac{3}{4} = 6\frac{5}{4} - 5\frac{3}{4} =$
$1\frac{2}{4} = 1\frac{1}{2}$

16. $9\frac{2}{3} + 6\frac{5}{9} = 9\frac{18}{27} + 6\frac{15}{27} =$
$15\frac{33}{27} = 16\frac{6}{27} = 16\frac{2}{9}$

17. $5\frac{1}{5} - 2\frac{5}{6} = 5\frac{6}{30} - 2\frac{25}{30} =$
$4\frac{36}{30} - 2\frac{25}{30} = 2\frac{11}{30}$

18. $2\frac{5}{8}$ in

19. $7X + 9 = 44$
$7X = 35$
$\frac{1}{7} \cdot 7X = 35 \cdot \frac{1}{7}$
$X = 5$

20. $7(5) + 9 = 44$
$35 + 9 = 44$
$44 = 44$

21. $\frac{3}{8}A - 8 = 13$
$\frac{3}{8}A = 21$
$\frac{8}{3} \cdot \frac{3}{8}A = 21 \cdot \frac{8}{3}$
$A = \frac{168}{3} = 56$

22. $\frac{3}{8}(56) - 8 = 13$
$21 - 8 = 13$
$13 = 13$

23. $\frac{5}{6}G + \frac{1}{6} = \frac{5}{12}$
$\frac{5}{6}G = \frac{3}{12} = \frac{1}{4}$
$\frac{6}{5} \cdot \frac{5}{6}G = \frac{1}{4} \cdot \frac{6}{5}$
$G = \frac{6}{20} = \frac{3}{10}$

24. $\frac{5}{6} \cdot \frac{3}{10} + \frac{1}{6} = \frac{5}{12}$
$\frac{15}{60} + \frac{1}{6} = \frac{5}{12}$
$\frac{1}{4} + \frac{1}{6} = \frac{5}{12}$
$\frac{5}{12} = \frac{5}{12}$

25. $\frac{\cancel{5}}{8} \times \frac{1}{\cancel{3}} \times \frac{\cancel{3}}{\cancel{5}} = \frac{1}{8}$

26. $\frac{\cancel{4}}{\cancel{5}} \times \frac{11}{\cancel{4}} \times \frac{\cancel{10}^{\,2}}{3} = \frac{22}{3} = 7\frac{1}{3}$

27. $\frac{4}{5} = \frac{80}{100} = .80 = 80\%$

28. $\frac{1}{4} = \frac{25}{100} = .25 = 25\%$

29. 15: $\underline{3}$, $\underline{5}$, $\underline{15}$
45: $\underline{3}$, $\underline{5}$, 9, $\underline{15}$, 45
GCF = 15

30. 2 x 2 x 2 x 7

31. $7\frac{2}{3} = \frac{23}{3}$

32. no

33. $\frac{22}{7}(21^2) = \frac{22}{\cancel{7}} \cdot \frac{\cancel{441}^{\,63}}{1} =$
$\frac{1386}{1} = 1{,}386$ sq. ft

34. $\frac{2}{1} \cdot \frac{22}{\not{7}} \cdot \frac{\not{21}^{3}}{1} = 132$ ft

Multistep Problems Lesson 6

1. area of garden is 25 x 15 = 375 sq ft
   $\frac{1}{3}$ x 375 = 125 sq ft for corn
   $\frac{1}{5}$ x 375 = 75 sq ft for peas
   125 + 75 = 200 sq ft used
   375 - 200 = 175 sq ft left over
2. $\frac{1}{2}$x 32 = 16 Sarah signed
   $\frac{3}{8}$x 32 = 12 Richard signed
   16 + 12 = 28 are signed
   32 - 28 = 4 cards left to be signed
3. $15 + $9 + $12 = $36 needed
   $\frac{1}{4}$ x $36 = $9 on hand
   $36 - $9 = $27 to earn

Multistep Problems Lesson 12

1. $\frac{1}{5}$x $30 = $6 for sandwich
   $\frac{1}{6}$x $30 = $5 for museum
   $\frac{1}{2}$x $30 = $15 for book
   $6 + $5 + $15 =$26 spent
   $30 - $26 = $4 left over
2. $\frac{1}{2}$x 128 = 64 miles Mark drove
   $\frac{1}{4}$x 128 = 32 miles Justin drove
   64 + 32 = 96 miles driven
   128 - 96 = 32 miles left
   32 ÷ 2 = 16 miles driven by Kaitlyn
3. $\frac{1}{2}+\frac{3}{4}=\frac{2}{4}+\frac{3}{4}=\frac{5}{4}$ pie
   $\frac{5}{4}\div\frac{1}{8}=\frac{40}{32}\div\frac{4}{32}=$
   40 ÷ 4 = 10 pieces of pie

Multistep Problems Lesson 18

1. 63 + 21 + 63 + 21 = 168 yd
   for outside of field
   21 + 21 = 42 yd for dividing sections
   168 + 42 = 210 yd of fencing
   210 x 3 = 630 ft of fencing
2. $\frac{1}{2}$ x 1824 = 912 with team colors
   $\frac{2}{3}$ x 912 = 608 with hats and colors
   $\frac{1}{4}$ x 608 = 152 with hats, colors, and banners
   152 + 100 + 25 = 277 with banners
3. 36"÷ 3 = 3 ft
   2 x 3 = 6 ft to brim, so cube is 6 ft on a side
   6 x 6 x 6 = 216 cu. ft volume
   216 x 56 = 12,096 pounds of water

Multistep Problems Lesson 24

1. $2\frac{1}{2}+1\frac{1}{8}+1\frac{1}{4}=4\frac{7}{8}$ pounds of salad
   $4\frac{7}{8}-1\frac{1}{2}=3\frac{3}{8}$ pounds left
   $3\frac{3}{8}\div 3=\frac{27}{8}\div\frac{3}{1}=\frac{27}{8}\times\frac{1}{3}=$
   $\frac{9}{8}=1\frac{1}{8}$ lb in a container
2. 1 x 1 x 2 = 2 cu. ft in aquarium
   $3\times\frac{1}{4}=\frac{3}{4}$ cu. ft for one kind of fish
   $2\times\frac{1}{3}=\frac{2}{3}$ cu. ft for other kind of fish
   $\frac{3}{4}+\frac{2}{3}=1\frac{5}{12}$ cu. ft for both kinds
   $2-1\frac{5}{12}=\frac{7}{12}$ cu. ft
   $\frac{2}{3}=\frac{8}{12}$, so there is not room for extra fish
3. $5\frac{1}{2}\div 1\frac{1}{2}=\frac{11}{2}\times\frac{2}{3}=\frac{22}{6}=3\frac{4}{6}=3\frac{2}{3}$ pieces
   $1\frac{1}{2}$ yd x 36 in = 54 in length of one piece
   $\frac{2}{3}$x 54 = 36 in length of left over piece

Symbols & Tables

SYMBOLS

=	equals
≈	approximately equal
<	less than
>	greater than
%	percent
π	pi (22/7 or 3.14)
r^2	r squared or r · r
'	foot
"	inch
∟	right angle

PERIMETER

Any figure –
add the lengths of all the sides

CIRCUMFERENCE

circle $C = 2\pi r$

AREA

rectangle, square, parallelogram:
$A = bh$ (base times height)

triangle: $A = \frac{bh}{2}$

circle: $A = \pi r^2$

trapezoid: $A = \frac{b_1 + b_2}{2} \times h$

(average of two bases times height)

MEASUREMENT

3 teaspoons (tsp) = 1 tablespoon (Tbls)
2 pints (pt) = 1 quart (qt)
8 pints (pt) = 1 gallon (gal)
4 quarts (qt) = 1 gallon (gal)
12 inches (in) = 1 foot (ft)
3 feet (ft) = 1 yard (yd)
5,280 feet (ft) = 1 mile (mi)
16 ounces (oz) = 1 pound (lb)
2,000 pounds (lb) = 1 ton
60 seconds = 1 minute
60 minutes = 1 hour
7 days = 1 week
365 days = 1 year
52 weeks = 1 year
12 months = 1 year
100 years = 1 century
1 dozen = 12

DIVISIBILITY RULES

Number is divisible by:

2	if it ends in an even number
3	if digits add to a multiple of 3
5	if it ends in 5 or 0
9	if digits add to a multiple of 9
10	if it ends in 0

VOLUME

rectangular solid: $V = Bh$
(area of base times height)
cube: $V = Bh$

EXPANDED NOTATION

1,452.5 = 1 x 1,000 + 4 x 100 + 5 x 50 + 2 x 1 + 5 x 1/10

Glossary

A

Additive inverse - a number that, when added to the original number, equals zero

Arabic numerals - the numerals used commonly today

Area - the number of square units in a rectangle or other two dimensional figure

Average - the result of adding a series of numbers and dividing by the number of items in the series

B–C

Base - the top or bottom side of a shape

Borrowing - see Regrouping

Canceling - when preparing to multiply or divide fractions, dividing a numerator and a denominator by the same number in order to simplify the problem

Carrying - see Regrouping

Century - one hundred years

Circumference - the distance around a circle, corresponds to perimeter

Coefficient - a number directly in front of an unknown. It is multiplied by the unknown.

Composite number - a number with factors other than one and itself

Cube - a three-dimensional figure with each side the same length

Cubic units - the result of multiplying three dimensions. Answers to volume problems are in cubic units.

D

Decimal or decimal fraction - a fraction written on one line by using a decimal point and place value

Denominator - the bottom number in a fraction. It tells how many total parts there are in the whole.

Dimension - the length of one of the sides of a rectangle or other shape

Divisor - the number that is being divided by in a division problem

E

Equation - a number sentence where the value of one side is equal to the value of the other side

Equivalent - having the same value

Estimation - used to get an approximate value of an answer

Even number - a number that ends in 0, 2, 4, 6, or 8. Even numbers are multiples of two.

Expanded notation - a way of writing numbers where each amount is multiplied by its place value

F–G

Factors - the two sides of a rectangle, or the numbers multiplied in a multiplication problem

Factor tree - a method for finding the prime factors of a given number

Fraction - one number written over another to show part of a whole. A fraction can also indicate division.

Greatest common factor (GCF) - the largest number that is a factor of two or more numbers

H–I

Height - the length of a line from the top to the bottom of a shape that forms a right angle with the base

Improper fraction - a fraction with a numerator larger than its denominator

Inequality - a number sentence where the value of one side is greater than the value of the other side

M–O

Mixed number - a number made up of a whole number and a fraction

Multiplicative inverse - a number that, when multiplied times the original number, equals one. Also referred to as the reciprocal

Numerator - the top number in a fraction. It tells how many of the parts of a whole have been chosen.

P

Percent - "per hundredth" - a way of writing fractions that have a denominator of 100. The symbol is %.

Pi - the Greek letter π, which has a value of approximately 22/7 or 3.14. It is used in finding the area or circumference of a circle.

Place value - the position of a number which tells what value it is assigned

Prime factors - all the factors of a number that are prime

Prime number - a number with only two factors, one and itself. (One is not a prime number.)

Product - the answer to a multiplication problem

Proper fraction - a fraction with a numerator smaller than its denominator

R

Radius - the distance from the center of a circle to its edge.

Reciprocal - a number in which the numerator and denominator have switched places. A number times its reciprocal equals one.

Rectangle - a shape with four "square corners," or right angles

Rectangular solid - a three dimensional shape with each side or face shaped like a rectangle

Reducing - dividing the numerator and denominator of a fraction by the same number. The resulting fraction is equivalent, but in "lower terms."

Regrouping - moving numbers from one place value to another in order to solve a problem. Also called "carrying" in addition and multiplication, and "borrowing" in subtraction.

Repeated division - a method for finding the prime factors of a given number

Right angle - a square corner (90° angle)

Roman numerals - a numbering system employed by the Roman Empire that uses letters to represent the numbers

Rounding - writing a number as its closest ten, hundred, etc. in order to estimate

Rule of four - a method for finding the same, or common, denominator of two fractions

S

Same difference theorem - an alternate method used to subtract mixed numbers and measures

Square - a rectangle with all four sides the same length

Square units - the result of multiplying two dimensions. Answers to area problems are in square units.

T–U

Triangle - a shape with three sides

Units - the first place value in the decimal system - also, the first three numbers in a large number. (starting from the right). The word units can also name measurements. Inches and feet are units of measure.

Unknown - a number whose value we do not know. It is usually represented by some letter of the alphabet.

V–Z

Volume - the number of cubic units in a three dimensional shape

Master Index for General Math

This index lists the levels at which main topics are presented in the instruction manuals for *Primer* through *Zeta*. For more detail, see the description of each level at www.mathusee.com. (Many of these topics are also reviewed in subsequent student books.)

Epsilon Index